Research Methodology and Quantitative Techniques

Research Methodology and Quantitative Techniques is a guide tailored for students and research scholars navigating the intricate landscape of research degrees across various disciplines.

From clearing coursework to formulating research synopses, selecting methodologies, conducting analysis and penning impactful theses, this book is a roadmap for every stage of the research journey. It empowers scholars to undertake original, quality research that not only fulfils academic requirements but also contributes to the burgeoning pool of knowledge in diverse fields. Uniquely structured to address the specific needs of researchers, this guide goes beyond traditional boundaries, delving into areas like IPRs and research ethics often overlooked in discipline-oriented texts. By offering comprehensive support, from topic selection to publication, it aims to be the go-to resource for researchers seeking a seamless path from inception to dissemination.

This book, *Research Methodology and Quantitative Techniques*, addresses every facet of research with clarity and insight and serves as both a companion and a vital tool for scholars poised to make a meaningful research impact in their fields.

K.A. Varghese is Statistician at Pacific Institute of Medical Sciences, Udaipur and retired Professor (Statistics), Maharana Pratap University of Agriculture and Technology, Udaipur.

B.R. Ranwah is a former ICAR Emeritus Scientist and retired Professor and Head of the Department of Genetics and Plant Breeding, Maharana Pratap University of Agriculture and Technology, Udaipur.

Nisha Varghese is Associate Professor at Indira Gandhi National Open University, New Delhi.

Nikhil Varghese is Associate Professor at Christ University, Lavasa Campus, Pune.

Research Methodology and Quantitative Techniques

A Guide for Interdisciplinary Research

K.A. Varghese, B.R. Ranwah, Nisha Varghese and Nikhil Varghese

Routledge
Taylor & Francis Group
LONDON AND NEW YORK

Designed cover image: sesame / Getty Images

First published 2025
by Routledge
4 Park Square, Milton Park, Abingdon, Oxon OX14 4RN

and by Routledge
605 Third Avenue, New York, NY 10158

Routledge is an imprint of the Taylor & Francis Group, an informa business

British Library Cataloguing-in-Publication Data
A catalogue record for this book is available from the British Library

ISBN: 9781032863597 (hbk)
ISBN: 9781032840338 (pbk)
ISBN: 9781003527183 (ebk)

DOI: 10.4324/9781003527183

Typeset in Times New Roman
by Apex CoVantage, LLC

Contents

Foreword

It gives me immense pleasure to know that a book titled *Research Methodology and Quantitative Techniques: A Guide for Interdisciplinary Research* written by very experienced former professors/faculty members of Maharana Pratap University of Agriculture and Technology, Udaipur and others is being published by Routledge. The importance of research in academic institutions has gone up tremendously during the recent past and the New Education Policy of the Government of India pinpoints the need to have quality research in Indian universities and colleges. The problem-oriented and location-specific research has been getting utmost priority in the ICAR system including State Agricultural Universities for the last more than four decades. With the UGC directives to clear the entrance test as well as research methodology courses by the scholars to get registered for research degrees from colleges and universities, the focus on research in various disciplines has been further streamlined. While student research in academic universities and colleges is only a partial fulfilment of degree programmes, it has tremendous scope to interlace with the institutional research output being generated by various research institutions in India. It will not only avoid duplication of research efforts but also raise the total research output to a greater level.

Academic research can be perceived as a science as well as an art since scientific methods are to be used for conducting research and choices among alternatives are to be made by scholars at various stages in research, like selection of research topic, research methods/designs, research hypothesis, quantitative techniques and so on. Besides, at the initial stage of research, the scholars with limited exposure to research find a lot of problems at different stages of their research. It is a fact that the books available on research methodology are discipline-oriented and confined to selected topics related to research methodology limiting the scope to set a strong platform by the researchers to carry out the research. The issues related to the role of research in the higher education system, the linkages of research with other functional areas of the university system, factors to be considered for the selection of topic, identification of appropriate research method and design, quantitative techniques to be used, ethical issues associated with research, publications, IPR related issues are becoming more and more pertinent in different disciplines like health and medical sciences, veterinary sciences, agriculture and others. Most scholars find it difficult to select a research topic in their respective disciplines, to formulate the research synopsis, to do a meaningful review work at the initial stages of research. The collection of data and its analysis using appropriate quantitative techniques are also very vital for the timely completion of a successful research by scholars in universities and colleges.

The lack of comprehensive reference books addressing all these issues has been a major problem for many researchers in various disciplines. The scholars have to refer to many books to prepare themselves to clear the research methodology course and to make a platform to pursue their research work on a scientific footing. I understand that this book being published by Prof. K.A. Varghese, Prof. B.R. Ranwah, Dr. Nisha Varghese and Dr. Nikhil Varghese is a comprehensive book on research methodology and quantitative techniques which will help the researchers right from selection of research topic to publication of research papers after completion of their research. It can be a good reference book for faculty members, research scholars and others engaged in the task of doing scientific research in various disciplines as the authors have different academic backgrounds in different disciplines.

I congratulate and compliment the authors and publishers for their great efforts in bringing out a comprehensive reference book on *Research Methodology and Quantitative Techniques* as one volume which can be a useful reference material and a guide for research scholars, faculty members, students and others across disciplines at various stages of their research.

Dr. N.S. Rathore
Former Deputy Director General (ICAR) Former Vice-Chancellor, MPUAT,
Udaipur and Former Vice-Chancellor, SKRAU, Jobner, and
Director, Geetanjali Institute of Technical Studies
Udaipur, Rajasthan, India

Preface

Research in higher education in all branches of learning has been getting increasing importance during the recent past. In fact, research and extension education/community services are integral parts of higher education in different branches of learning. Research, being a scientific and systematic probe which is more multidisciplinary in nature and the application of statistics at different stages of research has made it a difficult task for many scholars/students pursuing their post-graduate and PhD studies. As the majority of students in biological sciences come through the biological stream after the tenth class, it is natural for them to find subjects with mathematical applications difficult. In order to familiarize such students with research basics, online and offline courses in research methodology are offered by different institutions in higher education.

Realizing the fact that there is a shortage of comprehensive reference books on research methodology and there is a great need to have more reference books on research methodology including statistical applications specially designed for all streams of students, this book has been written by the authors. The book is designed in such a manner that the research scholars can identify their research topic, conduct it and do the analysis work themselves or with the help of a computer programmer/software expert.

The very purpose of writing this book is to encourage research scholars to do original quality research so that the huge research outcome through PG and PhD research forms part of knowledge in different disciplines/subject matter areas in different disciplines. Apart from part of partial fulfilment of respective degree programmes, the student's research must lead to the generation of new knowledge, theories, concepts, practices, etc. in different disciplines. Besides, student research can be a major part of location-specific problem solving as students can take up location-specific problems for their research. The active involvement of students at all stages of research will help to familiarize them with all issues in the conduct of meaningful research as a means for expanding the knowledge components in different subject matter areas of different disciplines.

The first three chapters are to familiarize students with research as a scientific activity. Chapter 3 would give enough background to prepare the synopsis or research protocol as the first step to conducting research. Chapter 4 gives the required knowledge on research methodology in general and research methods and designs widely applicable in different areas of study in particular which will help the students to identify the specific research methods and designs as applicable to the topic finalized by them. Chapters 5 and 6 give an outline of different designs available under experimental methods of research, mostly applicable in medical sciences and agricultural sciences. Chapter 7 on observational and epidemiological research covers the general research approaches for studies without interventions but based on field or laboratory observations. Chapters 8 outlines designs used for interview and survey methods including online surveys. Chapter 9 enlists various designs applicable to qualitative research.

Chapter 10 is meant to give exposure to research scholars on different designs applicable to sample selection. Chapter 11 outlines the scaling and scoring techniques applicable in research for the quantification of qualitative variables. Chapter 12 covers the concept of different types of research variables and research data. Chapters 13 to 17 cover the common statistical quantitative techniques used to analyze research data to generate results and findings of research. Chapter 18 is just to acquaint the students on the choices available for computer analysis of data including basic statistical analysis using options available under Microsoft Excel. Chapter 19 is on the documentation part related to synopsis, thesis and research papers. Chapter 20 is on how to write the bibliography and references at the end of the thesis and research paper as per the standards specified for it. Chapter 21 is to give a brief exposure to ethical issues in research and Chapter 22 is to familiarize researchers with IPR-based issues related to research.

While writing this book, the authors had to refer a large number of books, journals, research papers and other published and unpublished, online and offline sources including their own developed teaching material in different streams of education. We are highly grateful to authors and publishers of all such sources.

We have tried to briefly include various aspects of research methodology and statistical techniques. The book is intended to guide post-graduate and PhD students on all matters related to research as part of their degree programmes. The faculty members in different disciplines may also find it as a useful reference material. We aim to further improve and make this book a comprehensive one for students, faculty members and other researchers. Suggestions from all esteemed readers are most welcome.

1 Introduction

1.1 Introduction

Research in almost all branches of learning is closely associated with the development of the area and enhancement of the quality of life of people. Basic and applied research in various areas of study has made spectacular transformations in developing new theories, concepts, knowledge, technologies, etc. Continued research makes it possible to add new theories and information to the body of subject matter knowledge in various disciplines. Research focuses on widening knowledge on the one hand and aims for improvement and refinement of ongoing practices and techniques, search for newer techniques, interventions, processes, methods, etc. on the other hand. There are many definitions for research which is a systematic and scientific probe for new knowledge or an intensive search to find out solution to a problem.

Research and development (R&D) is very crucial in business, as it aims for useful knowledge which can pave the way for better business processes with increased production efficiency and reduced cost. It can also help businesses to have new products and services to meet the ensuing challenges from other players in the market. Research in all branches of learning not only keeps the subject matter vibrant but also adds to the body of subject matter knowledge. It is a tool for building knowledge and facilitates the learning process.

The word research is derived from the French term '*recherche*' or '*recerchier*' which means 'to go about seeking' or to look for something or to examine closely and carefully. Research can be defined as a systematic and scientific probe in any field for new knowledge, new information, new theory, new practice, new intervention, new invention, new product, new process or any such matters so as to answer a question or solve a problem. Research always aims to bring out something new. Research is a continuous process not only for the development of the subject matter but also for the development of the people as the output of research leads to enhanced quality of life of the people. Most of the institutional research aims to evolve new techniques and technologies which in turn enhance the quality of life of the people and thereby the mission of institutions gets popularized.

The close nexus of research with education and community/extension services makes the higher education system more vibrant. The problems of society are the determining factors for the research agenda in different areas of learning in higher education and the research output becomes a solution to their problems, and hence the social relevance of higher education is materialized. These days excellence, equity and expansion are the focal points of our higher education on the one side and on the other side, the system is striving for quality-oriented, job-linked and demand-driven higher education. Therefore, problem-oriented, location-specific and development-based research in our educational institutions is getting more and more emphasis. The penetration of academic professionals into local communities through extension education

DOI: 10.4324/9781003527183-1

or community services helps to get firsthand information about emerging location-specific researchable issues.

The complementary and supplementary association of institutional research with academic research made it possible for the subject matter areas of various disciplines and subjects to register spectacular growth in content and quality in areas of learning. The spillover effect of the research-based development transforms the socio-economic status of people the world over which is attributable to the evolution of need-based technologies in health, food, education, business and other associated sectors. The research-based innovations in both products and processes of these crucial sectors have paved the way to enhance quality improvement in the day-to-day lives of people. In the health sector, the quality and quantity of health services have increased tremendously leading to horizontal and vertical growth in institutional networks. In agriculture, health and business sectors revolutionary changes could be brought out through continued and systematic research. As a result, the magnitude of basic health indicators of people the world over has registered a rising trend. The need-based high-quality research for innovations and the people-friendly mechanism to transfer research outcomes across the borders of countries in the world altogether transformed the quality of life of the present people which is quite dynamic now as compared to what was available for just a few earlier generations.

Some of the definitions of research given in Research Methodology and Biostatistics—A Comprehensive Guide for Health Care Professionals are as follows:

> Research is defined as the creation of new knowledge and/or the use of existing knowledge in a creative way so as to generate new concepts, methodologies, and understanding.
>
> Research essentially is a problem-solving process, a systematic intensive study directed towards full scientific knowledge of subject studies.
>
> **—Ruth M. French**

> Research may be defined as the planned, systematic search for information for the purpose of increasing the total body of humanity's knowledge.
>
> **—Archold Lancaster**

In short, the definition of research is summarized in Figure 1.1.

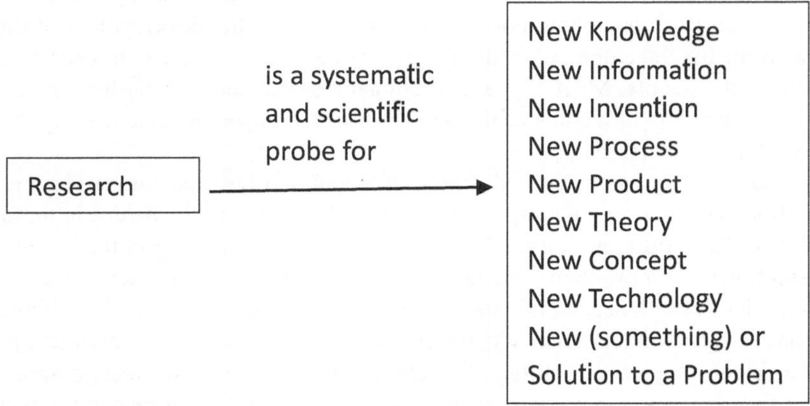

Figure 1.1 Definition of Research

Education in all spheres of study basically is a means for social development. The needs and aspirations of society continue to change with time. Higher education in all the branches of learning is moving towards an integrated approach of teaching, research and extension education/community services. Education becomes more effective and meaningful with such an approach. Integration of classroom teachings and hands-on practicals focused on different topics as per syllabus coupled with recent research outcomes and field problems related to the topics becomes more realistic for students in higher education. Therefore, education and research will have to move together so that regular updating and revision of pedagogy as warranted by time and supported by research is made effective in our educational system.

1.2 Types of Research

Based on the nature of the problem, purpose of study and research outcome, all the research fall under any one of the following two broad categories:

Basic/Fundamental Research: Basic and fundamental research is almost the same and aims to answer how things work so as to evolve new theoretical postulates related to the subject matter area of any subject. Basic research is more tedious, time-consuming and continuous in nature. Basic research is the way to enhance the scope of any subject.

Applied Research: Applied research is deductive in nature and aims to find solutions to emerging problems or find answers to a vital question related to an area of study. It can be the application of proven research results to solve a definite practical problem. It leads to generating new facts or information and sometimes has commercial applications.

Based on the approach, manner of conducting the research and outcome variables, the applied research can fall into one of the two following categories:

Quantitative Research is based on data (nominal, ordinal, discrete/countable or continuous/measurable) for both dependent (outcome) and independent (input) variables.

Qualitative Research relates to attributes or outputs qualitative in nature.

Over the years the ambit of research has been expanded and the categorization of research under different names based on the nature, scope, applications, method of conduct and so on has become more pertinent. Hence, several research types with a number of prefix or suffix terms have emerged over the years with different nomenclature. Some of these are:

Research with prefix of subject matter: The research as economics, social, agricultural, medical, epidemiological, surgical, homeopathic, ayurvedic, management, business, commercial, accounting, engineering, etc. is widely used these days to focus the subject matter relevance of research in respective areas.

Research with focus on place and agency of conduct: Research as institutional, inter-institutional, academic, college, departmental, faculty, students, industrial, collaborative, multi-locational, etc. is used to give emphasis as to where and how the research is managed or carried out.

Research-based on goals/objectives: Research as policy-oriented, problem-oriented, longitudinal, comparative, exploratory, descriptive, correlational, evaluation studies, impact studies, adaptive research, operational research etc. link to the aim or goal, purpose or broad objective of various such research work.

Research based on the manner of conduct: It includes quantitative and qualitative research types. Quantitative research as experimental, quasi-experimental, pre-experimental; as non-experimental such as observational, survey, descriptive, or analytical, case-control, cohort, meta-analysis etc.; qualitative research as phenomenological, ethnographic, grounded theory, case studies, historical studies, etc. based on the specific research methods and designs used to carry out various research studies.

Besides, there are some specific types of research reflecting the objective of research, some of which are:

Conceptual Research: Such research comes out with new concepts in any subject matter area. It enhances the scope of the subject matter of a subject.

Empirical Research: It comes out with empirical evidence related to new concepts evolved.

Diagnostic Research: It aims to assess the reasons for any happening as it is not apparently known earlier.

Clinical Research: It is focused on remedial measures for any unwanted or harmful occurrence of an event. It is also used to refer to research carried out in clinical units of medical colleges and hospitals.

Academic Research: The research carried out in higher academic institutions as a regular activity by faculty members, students or both.

Developmental Research: The research programmes for improving the quality of a product or a process and thereby the quality of life of the people falls under development research.

Action Research: Applied research for knowledge and socio-economic empowerment of people forms action research. Research and action go together in action research.

Cross-Sectional Research: Research carried out to assess the status of a situation at a point in time across various sections of society. It can be exploratory, descriptive or explanatory.

Longitudinal Research: It has a time dimension and is used to answer questions about temporal changes taking place over a specified time frame.

Time Series Study: The changes taking place over time in the values of a research variable with or without an intervention.

Panel Study: It is the combination of time series with cross-sectional study.

Cohort Studies: These are studies held on groups of homogenous subjects/objects to compare outcomes across groups including a control group.

1.3 Quality of a Researcher

A researcher must have certain qualities to be called a true researcher. A researcher must be creative with a critical mind and must have an analytical approach. He must have multi-disciplinary aptitude and be honest. The capacity to anticipate research output and to identify the real beneficiary of the research makes the researcher more focused.

1.4 Importance of Research

All research is development linked in one way or the other. The benefit of research is not only confined to the professional growth and development of the researcher but also to the society at large as most of the location-specific and problem-oriented research leads to the evolution of new techniques and technologies which is meant to improve the quality of life of the people or to reduce the burden of their life. Research which is a continued activity enhances the scope of the study area and field of study. It develops a critical mind for the researcher on the one hand and on the other hand, the mission of the institution where it is conducted gets popularized and brings name and fame for the researcher and the institutions.

1.5 Research Project

Any project is a time-bound activity with stipulated goals and targets. It will have stated costs and anticipated benefits. A research project is not an exception to these aspects of a project. The benefit of research is through the outcome of research. There are a number of sponsoring agencies to support the conduct of research as a project by competent implementing agencies. Sponsored research enhances the technical output of the sponsor, helps the implementing agencies to develop research infrastructure and research capability and the society to derive enhanced quality of life.

Some Important Points

- Research is a systematic and scientific probe for new knowledge, information, technology, etc. which adds to the subject matter content and enhances the quality of life of the people.
- The social relevance of research makes it a developmental activity for the welfare of the people.
- The broad types of research include basic and applied based on the goal and outcome of research.
- Based on the nature, the way of conduct and the outcome, the research can be quantitative or qualitative.
- Other categories of research are based on subject matter, place of conduct, manner of conduct, goals and objectives, research designs besides various other specific types.
- A researcher must be creative and honest with a critical mind and analytical approach.
- The benefit of research is not only confined to the professional growth and development of the researcher but also to the society at large as most of location specific and problem-oriented research leads to the evolution of new techniques and technologies which are meant to improve the quality of life of the people or to reduce the burden of their life.

Suggested Readings

Ghosh, B. N., *Scientific methods and social research*, Sterling Publishers Private Ltd., New Delhi, 1984.

Kerlinger, F. N., and H. B. Lee, *Foundation of behavioral research*, 4th ed., Harcourt College, Atworth, TX, 2000.

Sharma, B. S., *Research methods in social sciences*, Sterling Publishers Private Ltd., New Delhi, 1983.

Sharma, S. K., *Research methodology and biostatistics—A comprehensive guide for health care professionals*, Elsevier RELX India Pvt. Ltd., New Delhi, 2017.

Sidhu, K. S., *Methodology of research in education*, Sterling Publishers Private Ltd., New Delhi, 1985.

Trivedi, R. N., *Research methodology*, College Book Depot, Jaipur, 1991.

Williams, M. A., 'Editorial: assumptions in research', *Research in Nursing & Health* 3(2): 47–48, 1980.

2 Network of Institutional and Academic Research

2.1 Introduction

The total research output in different areas of study or operational areas comes from various exclusive research institutions as well as from academic institutions where research is an integral activity. Research institutes or research centres are established at international, national, state and regional levels by various governments and other agencies to focus research in various mandatory areas. Some institutes are meant for basic research while others focus on applied research. Technology-based economic development was given adequate importance ever since the planned economic development process was initiated in India after independence. A network of research institutions came into existence in India under various ministries and departments like the Indian Council of Agricultural Research (ICAR), Indian Council of Medical Research (ICMR), Council of Scientific and Industrial Research (CSIR), Defence Research and Development Organization (DRDO), Indian Council of Social Science Research (ICSSR), Indian Council of Forestry Research and Education (ICFRE), Indian Council of Historical Research (ICHR), Department of Science and Technology (DST), Department of Biotechnology (DBT) and so on. There are also several research institutions in areas like chemical sciences, physical sciences, mathematics, earth sciences, engineering sciences and material sciences, minerals and metallurgy, multi-disciplinary, etc. As technology-based development has been given prime importance in the development strategy in India, there are research centres in almost all areas of learning as well as related to production activities in India to augment the process of development in different areas of study and development.

2.2 Indian Council of Agricultural Research (ICAR)

The Indian Council of Agricultural Research (ICAR) is an autonomous body established to coordinate research and agricultural education in India under the Department of Agricultural Research and Education, Ministry of Agriculture. It is one of the world's largest networks of agricultural research and education institutes. The institutional network of ICAR includes four deemed universities, 65 ICAR institutions, 14 national research centres, six national bureaus and 13 directorates/project directorates, 63 state agricultural universities each having specific mandates for higher education and/or research. The ICAR also acts as a sponsoring agency for research projects to be implemented by various agencies.

2.3 Indian Council of Medical Research (ICMR)

One of the oldest medical research organizations in the world is the Indian Council of Medical Research (ICMR), which serves as the supreme authority in India for the formulation,

DOI: 10.4324/9781003527183-2

coordination and advancement of biomedical research. It is under the Department of Health Research, Ministry of Health and Family Welfare, Government of India. The ICMR established the Clinical Trials Registry - India in 2007 which serves as India's national registry for clinical trials. Research on specific health issues, such as AIDS, malaria, cholera, diarrheal diseases, vector control, nutrition, food and drug toxicology, reproduction, haematology, oncology, medical statistics, etc. is undertaken by the 26 national institutes of the ICMR. The other six regional medical research centres under ICMR focus on addressing region-specific health problems so as to strengthen and generate research capabilities in different geographic areas of the country.

Research priorities of the Council are in line with the national health priorities which include research on major non-communicable diseases like cancer, cardiovascular diseases, blindness, diabetes and other metabolic and haematological disorders; research on mental health; and research on drugs (including traditional remedies), control and management of communicable diseases, fertility control, maternal and child health, control of nutritional disorders and development of alternative strategies for health care delivery. The goals of these initiatives are to improve population health and well-being of the people while lowering the overall burden of disease. A variety of research initiatives conducted by medical students, faculty members and institutions are sponsored by the ICMR.

2.4 Indian Council of Social Science Research (ICSSR)

The Government of India established the Indian Council of Social Science Research (ICSSR) in 1969 with the goal of advancing social science research in the country. The ICSSR provides funding for projects, fellowships, international collaboration, capacity building, surveys, publications and other initiatives to enhance social science research in India. The ICSSR's documentation centre, the National Social Science Documentation Centre (NASSDOC), offers library and information support services to social science researchers. The ICSSR created the ICSSR Data Service to act as a national data service provider, enabling the social science community in India to share and reuse data, thereby fostering a robust research environment. The Division of Research Institutes and Regional Centres (RI&RC) manages regional centres around the nation and awards money for development and upkeep to research institutes. The Research Institutes are of all India character—outside the scope of the University Grants Commission (UGC). The main objectives of ICSSR are:

- Dispersal of talent from more developed to less developed regions, especially to areas where social science research is underdeveloped; and
- Development of quality of research and interdisciplinary research in social sciences to improve the social science inputs into development.

The Council is, at present, assisting 30 research institutes and six regional centres in different regions in India.

2.5 Council of Scientific and Industrial Research (CSIR)

The Council of Scientific and Industrial Research (CSIR) was established by the Government of India in September 1942 as an autonomous body that has emerged as one of the largest research and development organizations in India. It runs 38 laboratories/institutes, 39 outreach centres and three innovation centres throughout the nation. Although it is mainly funded by the Ministry of Science and Technology, it operates as an autonomous body through the Societies Registration

Act. The research and development activities of CSIR include aerospace engineering, structural engineering, ocean sciences, life sciences, metallurgy, chemicals, mining, food, petroleum, leather, environmental science, etc. In terms of Intellectual property, CSIR has a large number of patents in force internationally and nationally. The CSIR was awarded the National Intellectual Property (IP) Award 2018 in the category 'Top R&D Institution/Organization for Patents and Commercialization' by the Indian Patent Office.

2.6 Defence Research and Development Organization (DRDO)

The Defence Research and Development Organization (DRDO) is the premier agency under the Department of Defence Research and Development under the Ministry of Defence of the Government of India, charged with the military's research and development, headquartered in Delhi, India. It was formed in 1958 by the merger of the Technical Development Establishment and the Directorate of Technical Development and Production of the Indian Ordnance Factories with the Defence Science Organization. Subsequently, the Defence Research and Development Service (DRDS) was constituted in 1979 having Group 'A' Officers/Scientists directly under the administrative control of the Ministry of Defense. With a network of 52 laboratories, which are engaged in developing defence technologies covering various fields, like aeronautics, armaments, electronics, land combat engineering, life sciences, materials, missiles and naval systems, DRDO is one of the India's largest and most diverse research organization. The organization includes around 5,000 scientists belonging to the DRDS and many other subordinate scientific, technical and supporting personnel. As part of the rationalization plan, Defence Terrain Research Laboratory (DTRL) was merged with Snow and Avalanche Studies Establishment (SASE) which was renamed the Defence Geological Research Establishment (DGRE). Additionally, there are many institutional organizations in India under various departments of central and state governments or as autonomous bodies.

2.7 Other Research Organizations

There are many other research organizations in India focusing on research in specified areas. Some of these are:

- Indian Council of Forestry Research and Education (ICFRE)
- Indian Council of Historical Research (ICHR)
- Department of Science and Technology (DST)
- Department of Biotechnology (DBT)
- Indian Space Research Organization (ISRO)
- Central Power Research Institute (CPRI)
- Central Electronics Engineering Research Institute (CEERI)
- Central Food Technological Research Institute (CFTRI)
- Central Mechanical Engineering Institute (CMERI)
- Central Glass and Ceramic Research Institute (CGCRI)
- Marine Engineering and Research Institute
- National Botanical Research Institute

Apex academic bodies like UGC, AICTE and others also have provisions to support research in different areas. The need-based technologies developed by these institutions, either directly or through sponsored research projects through other institutions, not only add to the national knowledge bowl but also play a crucial role to ease the lives of the people by enhancing their quality of life.

2.8 Academic Research

India has the third largest public education system, after the USA and China, in the world with the University Grant Commission (UGC) as the apex body to regulate and monitor the higher education system. There are about 15 accreditation agencies approved by the UGC for quality assurance in various faculties of learning. As of 2020, India had about 1000 universities with 54 as central universities, 416 state universities, 125 deemed to be universities and 361 private universities. Additionally, there are about 160 institutes of national importance such as AIIMS, IIMs, IIITs, IISERs, IITs, NITs, etc. In all, there are about 53000 colleges in India. Every year nearly 25,000 PhDs are awarded and nearly 200,000 stand as enrolled for research degrees. Apart from central, state and private universities, there is a network of institutions offering PhD degrees in science, technology, agriculture, medicine and other areas under apex bodies like CSIR, ICSSR, ICAR, ICMR, etc. Thus, there are adequate opportunities for those who are desirous to do research in various subjects.

Due to the UGC regulations for PhD admission tests, regular coursework and its clearance, the PG and PhD research in academic institutions is a novel attempt in training for research methodology in different areas. The huge research efforts made by the educational institutions, teaching and research faculty members and also by the PhD scholars add to the research outcome of the country. The academic research output is yet to be fully recognized as a vital means for socio-economic development of society for improving their quality of life and also for the development of the subject in most of the areas. However, there are institutions focusing on problem-oriented and location-specific research which proved to be pivotal in adding new knowledge to the subject matter which is useful to society in one way or another. Hence the need of the hour is to interlace the research in higher education institutions with the research efforts carried out by the research institutions of high repute in different areas in the country as a means for development and quality enhancement of the life of ordinary people.

2.9 Integrated Teaching, Research and Extension Education/Community Services in Academic Institutions

Research has become an integral part of higher education in almost all branches of learning. With the advent of quality education in Colleges and Universities, the importance of research in higher education has increased tremendously. It gives the required training to students and scholars in higher education to formulate, implement, report and publish research outcomes so as to form a part of a major contribution to the national research outcome. The major parameters of quality education in higher educational institutions include updated curricula, effective delivery/teaching system, modernized infrastructure, vibrant faculty, adequate facilities for extracurricular activities, congenial campus environment, location-specific and problem-oriented research, time warranted and location-based community services/extension education and the strong interlinkage of teaching, research and community services. In other words, the teaching, research and community services/extension education in the higher educational institutions will have to become more and more integrated. This integrated approach supplements and complements each of these activities. The integrated model of teaching, research and extension/community services in the institutions for higher education is given in Figure 2.1.

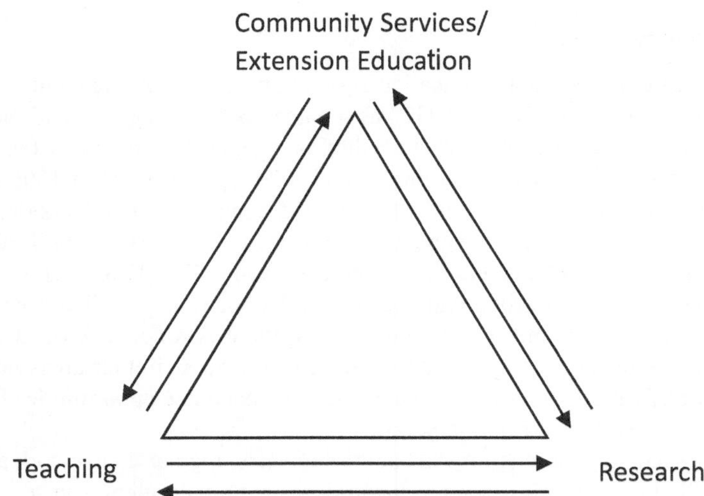

Figure 2.1 Integrated Model of Teaching, Research and Community Services/Extension Education in Higher Education

When a college teacher discharges all these functions that are complementary and supplementary to each other, the social relevance of higher education will be much higher on one side and on the other side the spectrum of higher education will be widened. For example, a teacher teaching a particular topic in a class based on contents in the latest edition of standard textbooks when linking his research and community service/extension education experience makes the teaching more effective. A teacher as a researcher must have the latest theoretical knowledge of the subject matter and the location-specific problems perceived through community services/extension education. Similarly, the extension services provided by the teacher as an extension specialist become perfect when he or she links theatrical knowledge with the latest research outcome related to the area of relevance. Over the years, the importance of integrated teaching, research and extension/community services is getting strengthened in India. In agriculture sciences, medical sciences, etc. the integrated teaching, research and extension/community services are relatively more in practice. With the idea of all colleges and universities adopting villages in rural areas, the cost-effective and quality of life-enhancing technologies will get popularized in rural areas.

The integrated teaching, research and clinical services/extension education in higher education institutions is comparable to a growing TREE where 'T' stands for effective teaching, 'R' for problem-oriented research and 'EE' for extension education or the community services that a college/university renders to the society.

A growing tree has many parts which broadly comprise the root system, the stem and branches including leaves. The root system makes the tree stand firm on the ground even at times of natural calamities and other adversities which is comparable to the teaching/education part of a university/college. A strong education system keeps the college growing over the years. The stem of a tree gives visibility to it from far-off places which is comparable to research in a university/college. A university/college doing good research will have visibility from far-off places through the quality of research-based publications it makes including publications by faculty members and students. The branches and leaves of a tree give shelter and shade to people, animals and birds and it is comparable to the community services and extension education it gives to the local community. Thus the 'TREE model' of higher education is going to be more relevant in the years to come where research plays a pivotal role in the growth and development of a university/college. The academic research conducted in universities and colleges along with the institutional research through the network of institutions in related areas makes the higher education system more vibrant and sustainable.

Some Important Points

In India, there are two broad systems of research in different areas. Firstly, institutional research is carried out on a regular basis in mandatory areas by the research institutes having specific research mandates. Secondly, the academic research carried out by PG and PhD students and faculties of higher educational institutes.

- The need of the hour is to interlink these parallel sets to avoid duplication and to make research more comprehensive.
- Integrated teaching, research and extension/community services in universities and colleges can make the higher education system more vibrant and effective.
- The teaching, research and community services/extension education in higher educational institutions will have to become more and more integrated.
- The 'TREE' model of higher education in India where T stands for Teaching, R stands for Research and EE stands for Extension Education/Community Services makes the higher education system more vibrant and effective.

Suggested Readings

www.csir.res.in
www.drdo.org
www.icar.org.in
www.icmr.org.in
www.icssr.org.in

3 Research Methodology, Research Topic and Research Synopsis

3.1 Introduction

Often the terms research methodology, research methods and research designs are used interchangeably. In the true sense, all these are different and each of these has a specific meaning. Research methodology is a wider concept in relation to research methods and research designs. In fact, research method is only a part of research methodology and research design is a part of research method.

There are two broad ways of doing research, namely institutional research and academic research. Those who do research as their main task in research institutions to meet its mandatory requirements have little option to choose their topics for research as the same is assigned to them in most cases. Secondly, those who do research as part of their academic activity or for partial fulfilment of the degree programme have a choice to select their research topic. The researchers in academic institutions are often confronted with the task of identifying a research topic, an appropriate research method and a suitable design as well as adopting suitable quantitative techniques befitting to a research problem. Hence a researcher must be aware of ways to select a topic, appropriate research methods and designs as well as analytical techniques so that he or she can choose the most appropriate methods and techniques to best resolve the challenges coming in the way of conducting the research. If the researcher has adequate knowledge of various research methods and analytical techniques, then the task of doing research becomes easy and interesting. Research methodology covers all those methods to identify a research topic and conduct the research to draw final conclusions using appropriate techniques including publication of research outcomes.

Most universities have designed courses on research methodology for PG and PhD students. The UGC has also made it mandatory for PhD students to undergo the course work and its clearance is necessary for registering as a PhD scholar in any department of the university or a college or institution. Research being a systematic process, the researcher must have knowledge about the complete process, from selecting a topic and making the synopsis to final submission of the thesis as well as publication of research results. In most universities, the research methodology (RM) course also includes modules on computer applications. The ethical considerations in research and research linked IPR issues also form part of RM courses in different universities.

Research methodology is a wider term as it involves aspects like the selection of a topic, review of literature, setting aims and objectives of research, fixing appropriate research method and design, determination of sample size and selection of sample, collection and compilation/computerization of data, data analysis, writing results, discussion, summary and conclusion, writing a bibliography, writing a thesis, publication of research papers, etc.

DOI: 10.4324/9781003527183-3

Hence a comprehensive research methodology course generally includes aspects like:

- Definition, types of research, institutional and academic research, research projects.
- Selection of a topic of research for academic research.
- Research synopsis, components and writing.
- Review of literature, sources, method of writing.
- Research methods and designs in different disciplines.
- Research variables and research data, types, sources, etc.
- Scaling, scoring and coding methods in research.
- Basic statistical methods and quantitative data analysis including descriptive and analytical techniques like measures of central tendency, measures of dispersion, correlation and regression analysis, concept of probability and probability distributions, sampling techniques, sample size, sampling distributions, statistical inference (parametric and non-parametric tests), multivariate statistical techniques, etc.
- Computer applications for data management and analysis in research.
- Ethical issues in research and publication.
- Intellectual property rights (IPR) and research.
- Technical writing and research publications.
- Bibliography and reference writing methods.

Efforts have been made to cover all these topics briefly in this book.

3.2 Stages of Academic Research

Research in any field or area is a scientific activity having specific standards and systematic procedures. The following stages of academic research are imperative to make research a systematic procedure.

(i) Planning

- Problem identification/topic finalization.
- Review of latest literature on the topic.
- Finalization of aim, objectives and research design.
- Synopsis drafting, submission and approval.

(ii) Execution

- Preparation of data collection format.
- Pre-test of data collection format.
- Sample selection.
- Pilot study.
- Revision of data collection format.
- Final data collection.
- Computerization of data.
- Validation of data.

(iii) Data Analysis

- Analysis of data for results (subject matter-wise and statistically).
- Compare results with results from related reviews.
- Draw inferences.

(iv) Drafting Report

- Prepare chapter scheme with major sections and sub-sections.
- Prepare draft report.
- Preliminary discussion with all concerned.
- Finalize the draft, report submission and paper publications.

3.3 Identification of a Research Problem

Identification of a research problem for academic research is a tedious task for a new researcher as it is a fresh area for many researchers. Location-specific and problem-oriented research is more relevant for academic research. Besides, research in academic institutions is time-bound as the entire work is to be completed within a stipulated time frame.

A research problem meeting the following aspects can prove to be a topic acceptable to the researcher as well as to the scientific community:

- Problem related to the mission of the academic institution.
- Searching solutions to an emerging problem/issue.
- Bridging the gap between 'what is' and 'what ought to be' for the area of research.
- Research problem convertible into research objectives.
- Must be clear, precise and self-explanatory.
- Can be expressed either as a question, a statement or a hypothesis.
- Scientifically probable.
- Leading to an analytical approach.
- Brings out new knowledge or information.
- Having anticipated benefit and beneficiary.

3.3.1 Selection of a Problem

The selection of a topic for research by a scholar in any higher educational institution can be done meaningfully with the following steps:

- Identify the research area of interest out of many areas of the subject.
- Extensive review of related work done in the identified area within the institution.
- Review of work done in the area of interest elsewhere.
- Visit websites of research organizations doing research on the same or similar areas.
- Close observation of field problems faced by people.
- Locate research gaps in the area and make out a tentative topic for research.
- Discussion with faculty members and others already doing research in the department.
- Listing and short listing of research problems.
- Finalize tentative aim and objectives of the identified research topic.
- Discuss the tentative topic with all concerned.
- One-to-one dialogue with major adviser/members of the advisory committee.
- Incorporate suggestions from peers and other researchers.
- Finalize tentative topic of research.

Thus, identification of a research problem is a tedious task for most of the researchers. The first and foremost thing that a researcher has to do is to develop a platform for his/her research. For that one should be very clear about the level of research done in the past by researchers in the same or similar topic elsewhere and in the same institution. In other words, the researcher must

Table 3.1 The FINER Concept

F	Feasible: For the researcher in terms of available time, required resources, self-competence, accessible guidance, availability of study units.
I	Interesting: to self, to the peers, professionals, society.
N	Novel: New, different from what is known or done already.
E	Ethical: Meeting the ethical requirements like informed consent from subjects, privacy of data, transparency, avoiding plagiarism, accuracy of data, etc.
R	Relevant: to the subject matter, society, target beneficiaries, etc.

be clear about the 'what is' of the topic. Then the researcher must ponder the question, 'what ought to be' about the topic. It will help the researcher to identify the real research gap. The topic of research so identified must be one searching solution to an emerging problem or bridging the knowledge gap. The problem must be convertible into specific research objectives which are clear, precise and self-explanatory. In other words, the research problem must be a scientifically probable statement expressed either as a question or as a hypothesis.

It is said that a research problem must be 'FINER' meaning that it must be—feasible, interesting, novel, ethical and relevant. Here, feasible means cost-effective, manageable, within the technical competence of the researcher and that can be completed timely. The research problem must not only be interesting to the researcher but also to peers and society as a whole. The novelty of the problem is in its uniqueness and innovativeness. The ethical aspect of research refers to the acceptance of the problem and the conduct of research in a manner acceptable to the scientific community. The research problem is relevant if its output contributes to scientific advancement and paves the way for further research (Table 3.1).

3.4 Research Synopsis

The primary step of every academic research is to develop a research synopsis or research protocol or research outline as the roadmap for research execution. A well-thought, thoroughly planned and drafted synopsis makes the task of the researcher easy and practically feasible. In order that any new research is a step forward in the field of study, the researcher has to make sure the terminal point of past research carried out by others in the same or similar areas which can be a platform to start research by the researcher. Hence a comprehensive review of past research work, right from the same institution to anywhere else, is of very high relevance. A meaningful synopsis will have the following items with some modifications in the nomenclature of these items and sequences as decided by the respective institutions.

- Title of research problem.
- Problem formulation/importance of the study.
- Aim and objectives.
- Review of literature.
- Material and methods.
- Plan of work.
- References.

A synopsis made in haste by the researcher will cause enormous problems during the course of conducting the research. Hence the time and energy put in by the researcher for preparation of a synopsis will ease the task in a big way leading to the successful completion in time. Normally, research scholars do not give that much emphasis in finalizing a meaningful synopsis and end up with a series of problems and hurdles.

The components of the research synopsis will have to be clearly understood by the researcher. Right from the topic identification to finalization of the draft synopsis, all aspects will have to be seriously treated and with the finalization of synopsis, the roadmap for research must be clear and ready for execution by the researcher. Academic research is always a time-bound activity. A crystal-clear synopsis makes it possible for the researcher to complete the task successfully in time. The half-hearted way of drafting the synopsis leads to delays in timely completion and sometimes leaves the task unaccomplished by the researcher forever.

The topic selection in most cases will have to be done by the researcher. In some cases, the research guide may assign or suggest a topic of his interest to the scholar. The sequence of steps in the selection of research topic by the researcher is given in Figure 3.1. The remaining sections in this chapter are devoted to explaining the components of a research synopsis so that a well thought out synopsis is prepared by the researcher.

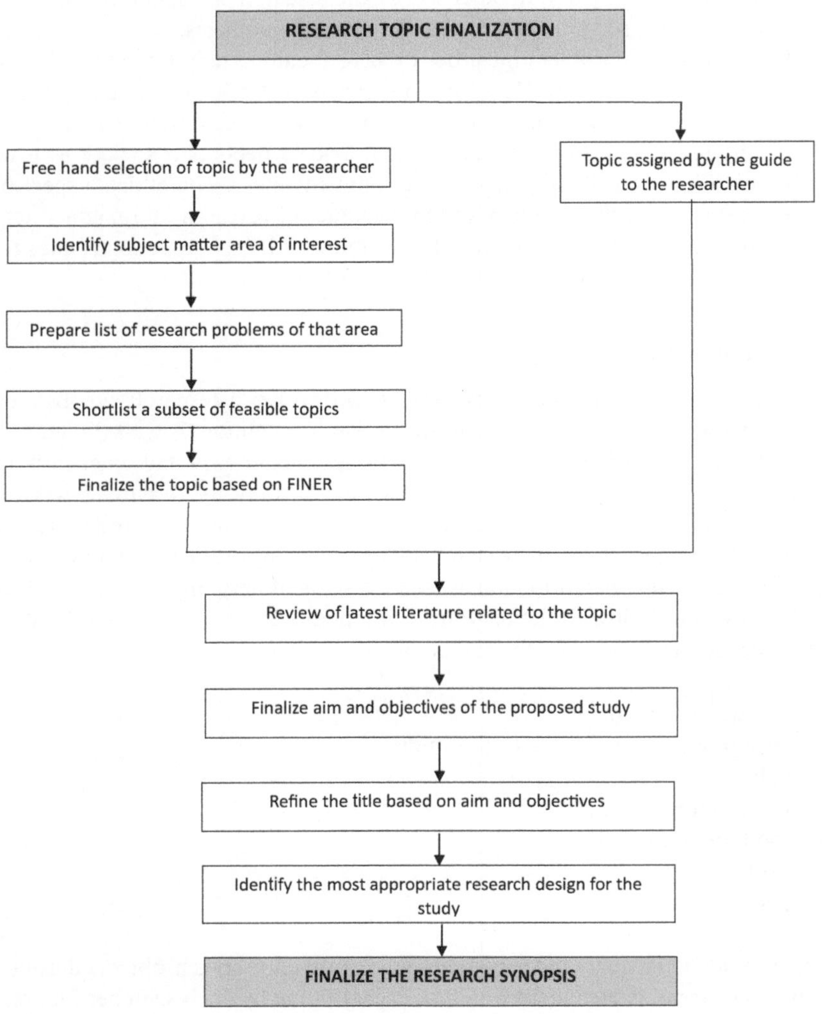

Figure 3.1 Sequence of Steps in Finalizing the Synopsis for Academic Research

3.5 Research Problem Formulation

It is the first technical section of a research synopsis and also termed as importance of the study of justification of the topic. Hence the researcher has to fully justify the importance of the topic. One can start with the origin/definition/emergence of the problem and its gravity at the global, national and local levels. The consequences and implications of the problem along with the efforts made to overcome the adverse implications and to promote positive implications can be stated. It would be useful if the current scenario is compared with the ideal situation so as to identify the research gaps. After enlisting all possible gaps, the researcher can identify the particular aspect being focused on by the researcher in the proposed study and its importance at different levels. The justification for the selection of the topic will have to be made clear in this section. The objectives, research hypothesis and expected outcome flow from this section of the synopsis.

3.6 Research Aim, Objectives and Hypothesis

The aim of research is the expected end goal of research in scientific terms. The research objectives of a topic include the split version of the scientific probe to achieve the aim that flows from the research problem itself. The following points are pertinent to research objectives.

- Flows from and linked to the topic.
- Limit the number of objectives (2–4).
- Normally starts with 'to study, to assess, to compare, to work out, to examine, to estimate'.
- Must be concise, precise and self-explanatory.

The research hypothesis restates the research aims as the expected concrete final outcome of research in a concise manner. It generally matches with the alternative statistical hypothesis. The aim is a concrete statement reflecting the expected final outcome of the research. The objectives reflect the strategies or supporting activities to arrive at a definite conclusion.

3.7 Review of Literature

It is the concise documentation of published research work related to a specific area/topic that the researcher has gone through and to be used at various stages of the research. The review of literature can be a good source to formulate/finalize the research problem, its objectives, methodology etc. In fact, it is the starting point of any research as the researcher has to ensure the present stage of research in the identified area related to the selected topic and to set a platform for further research. Hence, it is advisable to make searches for the latest reference materials from national and international published sources. Additionally, the researcher has to compare the findings of his/her research with those of other researchers in the discussion section of the thesis/report/paper and the review of literature plays a major role here. In short, the review of literature helps to make a stock of past work done, to shortlist related areas for further research, to identify a topic for research, to formulate the aim, objectives and research hypothesis, to identify data needs and sources, to acquaint with quantitative techniques to be used to address the stipulated objectives and above all to strengthen the content of research reports/thesis. Each review must be documented in a concise and precise manner.

Sources of review of literature include:

- Already completed research thesis in the same department/organization.
- Reports of completed/ongoing research projects.

- Published research papers in journals of the subject matter and related areas.
- Conference proceedings on topics related to selected areas/topics.
- Books/chapters in books published related to the area of study.
- Newspaper articles having relevance to the topic.
- Web sites of research organizations.
- Any other published or unpublished online or offline sources having authentic and presentable sources.

The researcher must do the review in a scientific manner by noting the following information in hard or soft copy to be used later on for the thesis chapters like introduction, review of literature, methodology, result and discussion and finally the bibliography to be given at the end.

- Author(s) name.
- Title of paper/book/report.
- Name of journal in the case of paper/title of the book.
- Volume and Issue number of the journal/publisher of the book and year of publication.
- Pages covered by the referred paper in the journal.
- A brief write-up covering the location of the study, the main objectives, methodology used and major outcome and conclusion of the study.

The documentation of the review of literature on these lines from the very beginning will help the researcher to write the report in the manner in which the reports/thesis/research papers are required to be written. In all cases, the bibliography/references are required to be given at the end of the thesis/paper in a specified style and the same is also based on reviews done by the researcher.

3.8 Material and Methods

The researcher has to define the population under study, the specific research design to be used, the method of sample selection with sample size, inclusion and exclusion criterion, the method and type of data to be collected, the quantitative techniques for data analysis, etc.

Research method refers to the manner of conducting the research to generate quantitative or qualitative evidence including ways to generate data required to address the stipulated objectives. Every research method has a number of specific designs to carry out research and generate data/evidence and the researcher has to select that design befitting to the specific situation or aim of the study. The proposed statistical and other quantitative techniques to cover the stipulated objectives may also be stated. The details of these research methods and research designs are discussed in the forthcoming chapters.

3.8.1 Research Method and Design

Research method is the method to generate data and evidence in support of the research hypothesis which is based on the expected outcome of research. There are various research methods and many designs under each method. The researcher will have to identify the best befitting method and design for the research problem. This book is meant to give exposure to researchers about the major research methods and designs applicable for different disciplines.

3.8.2 Sampling Plan

Normally every applied research is focused on a target population. The target population may not always be accessible to the researcher. Hence based on a defined sample population, a suitable sampling plan is used to select the sample for detailed investigation. The details of the sampling plan are outlined in Chapter 10. Based on the nature of the problem and target population, a suitable sampling plan will have to be identified and a sample of adequate size will have to be selected.

3.8.3 Inclusion Exclusion Criteria

In sample-based studies, the random sample selected based on the sampling frame (list of units in population) may include some units for which data collection may not be possible or inclusion of data of such typical units may cause disturbance or unnatural pattern in results estimated through sample observations. Hence the specific criteria for inclusion and exclusion of sampling units will have to be made at the time of sample selection and before the data collection is started so that the results represent the general/common population under study. This is more important in clinical, para-clinical and health science research. Even in other areas, such an approach is sometimes warranted to make results applicable to the population under study.

3.8.4 Data Collection

The quality of research data collected through the identified research method and design has a great role in the accuracy of results obtained in any research. Hence utmost care and caution will have to be made to collect the true and correct data from the respondents or units of the selected study. The sampling errors, due to the fact that the study is sample-based, as part of the larger population under study can be minimized by having the appropriate sampling plan and sample size for the study. The non-sampling errors which are due to rectifiable causes will have to be minimized by using the most befitting scientific methods for data collection. In experimental research, there may be standard equipment to record data of the subject of study at the required point in time. In opinion and perception-based survey methods, the data or opinion may not be always consistent over time even for the same respondent. In observational studies also the time reference of data is very vital. Hence the likely non-sampling errors in data will have to be minimized at the time of data collection. In all methods of research, there should be a pre-designed data format for a pre-fixed time reference for data collection which is guided by the objectives of the study. The data collection format may be prepared in such a way that all required data to fulfil the stipulated objectives are collected from each of the respondents and no unwanted data is collected.

3.8.5 Data Analysis

Broad statistical quantitative techniques include classification of data, tabulation of data, graphical/diagrammatic presentation of data, measures of central tendency (mean, median mode, quartiles, deciles, percentiles), measures of dispersion (standard deviation, mean deviation, range), measures of skewness (asymmetry), measures of kurtosis (flatness or peakness), correlation and regression analysis (associations and relationships of variables), Analysis of variance (in case of multiple samples), estimation (population parameters are estimated as sample statistic), testing of hypothesis (parametric and non-parametric) and so on.

The application of appropriate research methods for data collection and quantitative techniques for data analysis is very pertinent to draw meaningful inferences and conclusions from research. Every subject matter has its own measurable indices, indicators and techniques. In research, one can also make use of such measurements of other related subjects. In multi-disciplinary research, it is very much required to use such concepts and measurements of related subjects. The right selection of quantitative analytical techniques befitting the scope and objectives of the study is quite vital.

Besides, statistics is such a subject that its quantitative techniques can be applied in the research of many of the disciplines. Statistical techniques for descriptive and analytical objectives of various research studies are available. Every researcher may not be well-versed with all such techniques. Hence it is always better for the researcher to consult the statistician at different stages of the study. It is in the fitness of things to consult the statistician at the time of preparation of synopsis for purposes like selection of sampling method, sample size, analytical techniques for different objectives, etc. The broad type of data analysis using univariate, bivariate and multivariate statistical techniques is summarized in Figure 3.2.

The analysis of data can be perceived under three broad heads:

(i) Analytical techniques related to the main subject matter area of the topic.
(ii) Analytical techniques/quantitative methods from statistical methods.
(iii) Analytical techniques from related subject matter of the topic.

The knowledge of various quantitative techniques helps the researchers to apply the most appropriate techniques in different situations which in turn helps the researcher to come out with meaningful inferences. In fact, the quantitative techniques must function in a cafeteria mode so that there exists a proper interface between research problems and quantitative techniques. Knowledge of various quantitative techniques will help the researcher to use the most appropriate technique suitably to address the research objectives and also to arrive at concrete conclusions.

In univariate studies, the measures of central tendency like mean, median, mode, etc. help to describe the phenomenon under study for a meaningful understanding. In some cases, the variability in the values can be of great importance. The range, mean deviation, standard deviation, coefficient of variation, etc. makes it possible to assess variability or relative variability in the

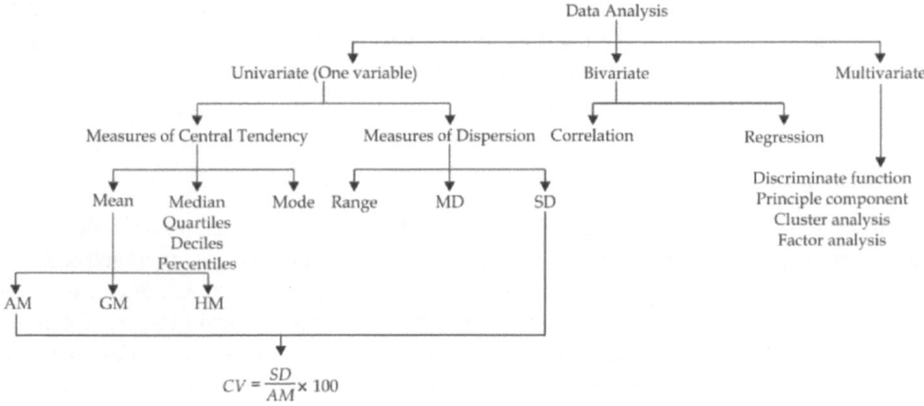

Figure 3.2 Options for Statistical Data Analysis

set of data under analysis. The coefficient of variation is a suitable measure to assess relative variability in different sets of comparable classes.

While correlation studies are used to assess the extent of association in a related set of data in bivariate/multivariate studies, regression studies are helpful to measure the impact of the independent variable on dependent variables in functional forms, especially for prediction purposes.

When we have a set of observations on each unit of study and the same set of information for many such units is available, it is possible to use multivariate studies, and a good number of such techniques are available. In system-based studies multivariate analysis is possible.

The clarity of perception of the researcher on the research topic, objectives of the study, research method and design, data coverage, quantitative techniques to be used for data analysis will not only make academic research a systematic and scientific probe but also make the task easier and interesting for the researcher. As research is a systematic and scientific method, it warrants a certain sequence of action so as to complete the task successfully within the stipulated time.

The statistical quantitative techniques commonly used in research are briefly discussed in Chapters 13 to 17.

3.9 Major Stages in Academic Research

The sequence of actions and major stages in academic research is summarized (Figure 3.3).

The subsequent chapters are meant to discuss various research methods and designs and also the statistical quantitative techniques generally used in academic research.

Figure 3.3 Sequence of Action in Academic Research

Some Important Points

- The research methodology course designed for PG and PhD students in universities and research institutes is to prepare a scholar for systematic conduct of research.
- The broad stages in research include planning, execution, data analysis and reporting.
- The selection of topic for PG and PhD research on 'FINER' (feasibility, interesting, novel, ethical and relevant) approach can make the research more practical, meaningful and for timely completion.
- A systematically prepared synopsis makes the task easier and simpler for the researchers.
- Systematic review of literature can help the researcher from problem formulation/identification to thesis writing and research publications.
- The research methodology course is a prerequisite for doing good research and includes concepts, types, designs, selection of topics, review related issues, synopsis preparation, stages of research, quantitative techniques in research, ethical issues, etc.
- It also includes basic statistical methods and quantitative data analysis including descriptive and analytical techniques.
- It also includes computer applications for data management and analysis in research.
- Besides, ethical issues in research and publication, bibliography and references writing methods, inclusion-exclusion criteria in sample-based studies are also part of RM.
- The quality of research data collected through the identified research method and design has a great role in the accuracy of results obtained in any research.
- The sampling errors can be minimized by having the appropriate sampling plan and sample size for the study.
- The non-sampling errors will have to be minimized by using the most befitting scientific methods for data collection.
- The application of appropriate methods for data collection and quantitative techniques for data analysis is very pertinent to draw meaningful inferences and conclusions from research.
- Every subject matter has its own measurable indices, indicators and techniques. In research, one can make use of such measurements of other subjects including various statistical techniques.

Suggested Readings

Bruce, C. S., 'Supervising literature reviews', in *Quality in postgraduate education*, edited by O. Zuber-Skerritt and Y. Ryan (eds.), Kogan Page, London, 1994.

Cooper, H. M., *Integrating research: a guide for literature reviews*, 2nd ed., Sage Publications, Newbury Park, CA, 1989.

Ghosh, B. N., *Scientific methods and social research*, Sterling Publishers Private Ltd., New Delhi, 1984.

Hart, C., *Doing a literature review*, Sage Publications, London, 1998.

Kerlinger, F. N., and H. B. Lee, *Foundation of behavioral research*, 4th ed., Harcourt College, Atworth, TX, 2000.

Leedy, P. D., *Practical research: planning and design*, 6th ed., Merrill, Upper Saddle River, NJ, 1997.

Sharma, B. S., *Research methods in social sciences*, Sterling Publishers Private Ltd., New Delhi, 1983.

Trivedi, R. N., *Research methodology*, College Book Depot, Jaipur, 1991.

4 Research Methods and Designs

4.1 Research Methods and Designs

Research is a science as well as an art. It is a science as scientific methods are used in the conduct of research. It is an art as the selection out of all available alternatives has to be made by the researcher at various stages of research right from the selection of topic, selection of appropriate research method and design, selection of most befitting quantitative technique and so on. Most of the academic research in applied areas is based on evidence or information collected from selected respondents or subjects of study selected from a defined population. This information can be observations or opinions collected by the researcher from the selected respondents. Research method is a part of research methodology and research designs are a subset of research methods.

Research methods are the ways of conducting research for generating data, observations and evidence as per stipulated aims, objectives and hypothesis of the research topic and in support of results/findings and conclusions/recommendations emerge after the study is completed. While the basic research leads to the formation of new knowledge as theories, principles, concepts, etc. is a continuous process, the applied research is mostly problem-oriented, location-specific and time-bound which can be quantitative or qualitative in nature. The methods in applied research aim to generate data based on responses, observations, opinions, etc. in respect of the respondents or subjects selected from the study population. The broad classifications of research methods on the basis of manner of conduct and the outcome of research are:

- Quantitative methods (measurable outcome).
- Qualitative methods (outcome as description).
- Mixed methods (combination of quantitative and qualitative).

  - **Quantitative research** is based on the collection and analysis of quantifiable data and focuses on empirical evidence or numerical outcomes. It generally follows a deductive approach. A range of quantifying research methods like interventional or experimentation on subjects of study, non-interventional observations recorded from the subjects of study, survey and interview responses as quantifiable perceptions, experiences, feelings, opinions, etc. from selected respondents and other goal-specific methods with measurable outcomes are widely used under quantitative research. The outcomes of quantitative research are generally in numerical forms and values.
  - **Qualitative research** involves the collection and analysis of non-numerical information (text, video, audio, dialogue, etc.) to understand perceptions, experiences, feelings, etc. of the subjects of study. It provides in-depth insights of the people on a problem or an issue. It is generally used for research in humanities, social sciences, psychology, anthropology,

DOI: 10.4324/9781003527183-4

sociology, education, history, etc. The data collection methods in qualitative research include direct observations, interviews, focus group discussion (participatory rural appraisal- PRA; rapid rural appraisal-RRA and so on.), surveys with open-ended questions and secondary sources like text images, audio, video recordings, etc. Besides, online research is carried out in areas like business, marketing and other areas. The outcome of qualitative research is generally in the form of description in text form.

- **Mixed methods** are the combination of both quantitative and qualitative methods.

4.2 Quantitative Research Methods

Some of the generally used quantitative research methods on the basis of conduct of research to generate data or evidence in support of research outcome are:

- Experimental/interventional research method (mostly used in agriculture, health/clinical sciences, veterinary, education, psychology and many other areas).
- Non-experimental/non-interventional observation research method (mostly used in social sciences, humanities, commerce, para-clinical, epidemiology, psychology, sociology, economics and many other areas).
- Survey and interview research method (used in social sciences, humanities, education, commerce, marketing, business and many other areas).
- Goal/situation based specific methods (used in all areas of learning).

Each of these methods has a number of different research designs. Research design is a specific way of conducting research under different methods or situations as per the stipulated aim and objectives of the study. A particular research problem can be carried out by applying different research methods and designs, but for each problem, there can be one most appropriate method and design. Hence the knowledge of various research methods and designs helps the researcher to identify the best method and design to be used for the specific problem under a specific situation. The selection of a research method and design depends on the nature and size of the population under study, the nature of the research problem, the goal of the study, researcher's exposure to different research methods and designs, ethical issues, type of participants and availability, access to required material, requirement of monetary and manpower resources for the study, time available to complete the study and the extent of control that the researcher can have on extraneous factors and many others. For some of the research problems, alternative research methods and designs are possible. In any case, there is always a best-fit research method and design for a specific research problem so as to ensure the authenticity and precision of research results.

4.3 Experimental/Interventional Method

An experiment is a procedure used to accept or reject a research hypothesis on the basis of evidence given by the outcome/dependent variable due to the intervention or manipulation made on a particular factor or factors of the study. In experimental research normally there will be a control so as to objectively compare, assess and quantify the impact or response of any specific intervention or manipulation on the subjects of study. Both the experimental and control groups are drawn randomly from the same population. It is generally applied in agriculture, education, health and clinical research, biological sciences, etc. where one or more interventions (as treatment/manipulation/addition/alteration) are made by the researcher on the subjects of study to assess the corresponding impact on the outcome variable of the study. The interventions can be a new method of teaching, a new crop variety, a new practice, a new drug/fertilizer, a new combination of drugs/fertilizers, a new dose of a drug/fertilizers, withdrawal of any drug or combinations, a new method of administering the inputs or drugs, a new method or process of treatment or surgery, training, counselling and so on.

Validity of a research design: The authenticity of research results depends on the research methods and design chosen for a research study. The validity of research results can be evaluated by using internal and external criteria.

The internal validity ensures whether there is a real difference in the outcome variables (interventional effect) in the experimental group compared to control group or the difference is due to some extraneous factors. If the experiment is able to establish that the effect (on dependent/outcome variable) is due to the independent variable (treatment or intervention), then we say that the experiment has internal validity. Randomization either in the formation of experimental and control groups or in the application of treatments randomly to the units of study is the technique to ensure internal validity of results. The extraneous factors that can disturb the internal validity of results include proxy factors of treatment (irrigation in fertilizer trials), the psychological effect of pre-test results on post-test values, change in instruments and measurement methods for pre and post-tests, replacement of dropouts in the study, non-homogeneity of populations (treatment and control groups), non-random selection of subjects for the study and similar factors can adversely affect the internal validity of results.

The external validity refers to the application of research findings to an extended population or the scope for generalization of specific research results to other subjects in the population. The factors contributing to external validity include the Hawthorne effect (giving positive responses as a member of a study group), researcher-respondent relationship, pre-test knowledge effect, socio-cultural differences of people, geographical effect and temporal effect and so on.

4.3.1 Research Designs Under Experimental Research

There are many research designs applicable under experimental research methods in areas like agriculture, health sciences, education, psychology, business, etc. These designs are:

- True experiment design (TED) (post-test only design, pre-test post-test design, Solomon four-group design).
- Randomized control trial (RCT).
- Completely randomized design (CRD).
- Randomized block design (RBD).
- Latin square design (LSD).
- Factorial design (FD).
- Quasi experimental design (QED).
  (Non-randomized control group design, time series design).
- Pre-experimental design (PED).
  (One-shot test design, one-shot pre-test post-test design).

More details about these designs are given in Chapter 5 and Chapter 6, respectively.

4.4 Non-Interventional Observation Methods

Non-interventional observation research is a major and widely used research method in social sciences, business, commerce, education, health sciences, psychology, agriculture, etc. In medical sciences, it is applicable in both clinical and non-clinical areas. The research in basic medical science subjects (anatomy, physiology, biochemistry, etc.) and community-based research in areas of community medicine, are mostly based on non-interventional observations only. Researchers make no external or internal manipulations to subjects of study as part of non-interventional observation research. It is widely used to assess the status of a situation in a population/system or to correlate internal factors or independent variables of the population/system. In fact, there are situations where independent variables cannot be manipulated due to ethical or other reasons and non-intervention observational study remains the best option in such cases. Natural occurrence of events or the observations made through laboratory or field investigations without any interventions or any deliberate control of events in any form is the basis of non-interventional observation research. In such research, observations are made without any manipulation of independent variables. The observations/data is collected through suitably structured data format based on personal observation, laboratory tests, medical test reports; OPD/IPD records etc. which are analyzed to have findings to arrive at a conclusion. There are specific designs befitting different situations under the non-interventional observation method of research. There is wide application of this method in applied areas. In these areas, the observations/data are generally collected from a selected sample of the population. Observational studies, while providing the status of a situation or relationship among variables under study, also pave the way for further research by developing meaningful hypotheses. The observational studies may include physical observations (personal or field), clinical observations (physical or laboratory), epidemiological observations (community-based) and interview-based responses. Such studies are mostly cross-sectional in nature and can be descriptive or analytical based on the objectives of the research.

4.4.1 Non-Interventional Observation Research Designs

The observations in numerical form, either physical or values, are widely used in applied research in many areas like agriculture, health sciences, economics, sociology, commerce, business, education and many others. Non-interventional studies are carried out either to describe a situation or to come out with cause-effect relationships in the system. Generally random samples are drawn from the population under study to collect information to estimate population parameters or relationships.

Most of the research in para-clinical sciences like anatomy, physiology, biochemistry, microbiology, pathology, forensic medicine, community medicine, etc. and also some studies in clinical subjects are based on observations related to patients as recorded in OPD, IPD, laboratories and operation theatres and even after hospital discharge. The observed changes in physiological functions, anatomical factors and pathological and biochemical parameters in relation to various morbidities/disabilities are the basis of research in these areas. Most of the non-interventional observation research is descriptive or analytical in nature. These designs are used to describe incidence, prevalence, causes, inter-relationship of various factors associated with diseases or events. It can also be used to describe the frequency of occurrence of a phenomenon.

The research designs under non-interventional observation research methods are:

- Descriptive case reports.
- Descriptive case series.
- Descriptive cross-sectional studies.
- Descriptive longitudinal studies.

- Analytical cross-sectional studies.
- Analytical longitudinal studies.

More details on different observational research designs are given in Chapter 7.

4.4.2 Epidemiological Research Methods

Epidemiological research forms a major area of research in health sciences. Epidemiology is the study of distribution (with respect to people, place and time) and determinants of diseases. Both interventional and non-interventional research designs are used in epidemiological studies. The interventional research designs have been listed in the previous section. The non-interventional observational research designs used in the epidemiological category of studies are:

- Case-control research.
- Cohort studies: A cohort is a group of homogenous individuals with respect to a phenomenon which can be a case or exposure to a risk factor. Cohort studies include:

  Prospective cohort
  Retrospective cohort
  Ambispective cohort studies

- Field trials.
- Community trials.
- Uncontrolled natural trials.
- Natural experiments.
- Before and after comparison trials.

More details on epidemiological research designs are given in chapter 7.

4.5 Survey, Interview and Online Research Methods

In surveys, interviews and online research we generally collect quantitative or qualitative data on the status, perception, experience, feeling, expectations, knowledge, practice, attitude, etc. of the respondents on a specific selected aspect. Surveys, interviews and online research play a crucial role in areas like business, marketing, customer satisfaction, commerce, social sciences, impact studies, KAP (Knowledge, Attitude and Practice) studies, adoption rates of technologies, descriptive and analytical cross-sectional studies and many others. Such studies can be complete enumeration (census-based) for small-size populations, random sample for large populations and purposive sample-based studies wherever random sample selection is not possible due to one reason or the other.

4.5.1 Survey Method

A survey in the true sense is the process of collecting, analyzing and interpreting data from many individuals/respondents with the aim to determine insights, perceptions and opinions from a selected group of people about a specific item/event/product, etc. A survey goes much deeper than a questionnaire and can involve more than one form of data collection. A survey is a combination of questions, processes and methodologies that analyze data collected from the participants. Most of the surveys involve questionnaires. The ultimate purpose of a survey is to find out more about the opinions, insights and perceptions of a certain group of people on a specified aspect. This is done for many reasons. For example, business surveys are used to find out more about consumer behaviour. A single questionnaire is only one small part of a survey. Otherwise, the approach for both survey and interview methods is more or less the same. Survey research has different modes for conducting it i.e. e-mails, online, telephonic, face-to-face, etc.

The survey research has two broad designs:

- Cross-sectional survey research.
- Longitudinal survey research.

4.5.2 Interview Method

Survey research and interview research are often used as synonymous with each other. But the basic difference is that for interview research the direct or indirect personal contact between the researchers or their representative with the respondents will be there, but for survey-based research, it is not necessary. A questionnaire is generally used for interviews. The purpose of the questionnaire is to gather data from a target audience on specific items in a systematic manner. It will include open-ended questions, closed-ended questions, multiple-choice questions, etc.

In interview research, the investigator will have to move directly or indirectly from one subject of study to the other to get the required data through direct or indirect contacts whereas in survey-based research the required items of information are listed in the form of a questionnaire and can be collected directly or indirectly from the respondents. Both methods are applicable for quantitative and qualitative research. A variety of studies can be conducted using survey and interview methods. These studies are possible for census/population-based studies, random sample-based studies and non-random sample-based studies.

The interview-based research can be carried out under different modes or designs. These include:

- Questionnaire-based indirect interview.
- Schedule-based direct interview.
- Internet-based Google form.
- Telephonic interview.

Such interviews can be structured, semi-structured or unstructured. These interviews can be of the following types:

- Personal interview.
- Telephonic interview.
- Email or web page interview.

4.5.3 Online Research Methods and Techniques

Researchers can collect data from the respondents using various online research techniques. They are often called internet research or web-based research methods. Many of these research methods are already being used in one way or the other but are being revived for the online mediums. The latest in this type of online research method is social media research as it offers extended levels of complexity and thus, new avenues for research are created. Both quantitative and qualitative research can be carried out using online options. Online research methods broadly include the following:

- Online focus group.
- Online interview.

- Online qualitative research.
- Online text analysis.
- Social network analysis.

Online marketing and business research covers aspects like customer satisfaction, new product response tests, brand loyalty, employee satisfaction, etc. Under situations like the Covid-19 period, online research was widely used.

More details on survey, interview and online research designs are discussed in Chapter 8.

4.6 Other Quantitative Research Methods

There are other research methods based on specific purposes, objectives and situations. Some of these are:

Methodological Studies: Studies aimed to find out subject matter-specific approaches, measurements, techniques, etc.

Meta-Analysis: Studies by pooling the results of many similar studies conducted at many locations for wider applications.

Evaluation Studies: Studies conducted to evaluate already implemented programmes and policies.

Operational Research: It is the application of scientific methods of investigations to study the complex human organization and services. It relates to studies to generate macro level evidence based on micro level research results.

Knowledge, Attitude and Practice (KAP studies): KAP studies are generally applied to assess the status of any phenomenon having relevance to the people. The level of knowledge and the attitude of the people are determining factors to the extent of practice. Such studies are carried out using well-formulated questionnaires to quantify the level of knowledge, to assess the attitude of the people and also the extent of practice followed by the subjects of the study phenomenon. Usually, Likert scale-based scores are developed for each respondent. Then it is possible to examine the association of knowledge and attitude with respect to practice.

Implementation Research: It is defined as research on knowledge linked to medical practice for better health of a community. It is a participatory form of research in medical sciences where all stakeholders of health sciences join hands with each other at different stages from planning, implementation and evaluation and action plan for the benefit of the community.

Translational Research: Translational research also known as 'Bench to Bedside Research' is the process by which the results of research done in the laboratory are directly used to develop new ways to treat patients. It advocates the need to interface the clinical sciences and basic sciences. It has emerged as a new interdisciplinary branch of medical sciences. It aims to translate the findings of basic fundamental research into medical practice and meaningful health outcomes. It is an interdisciplinary branch of medical sciences which integrates bench side, bedside and community research. Findings from basic science research are applied to human health and medical practices.

Despite contextual differences between research methods and research designs, these terms are interchangeably used by the researchers. What matters is the right interface between the research problem and research method so as to scientifically generalize the evidence given by the sample to the population under study. The broad research methods and different research designs under each method are summarized in Table 4.1.

Table 4.1 Broad Quantitative Research Methods and Designs Under Each Method

Quantitative Research Methods and Designs						
Experimental/Interventional			*Non- Experimental/ Observation*	*Epidemiological*	*Interview/Survey and Online*	*Other Methods*
True Experiment	*Quasi-Experimental*	*Pre-Experimental*				
i. True Experiment ii. RCT iii. Post-test iv. Pre-test post-test v. Solomon four-group vi. CRD vii. RBD viii. LSD ix. Factorial	i. Non-Random Control ii. No Control Group	i. One Group Pre-test Post-test	i. Descriptive Case Report ii. Descriptive Case Series iii. Descriptive Cross-sectional iv. Analytical Cross-sectional v. Descriptive Longitudinal vi. Analytical Longitudinal	i. Case Control ii. Prospective Cohort iii. Retrospective Cohort iv. Ambispective Cohort v. Field Trials vi. Community Trials vii. Uncontrolled Trials viii. Natural Experimental ix. Before After Comparison Trial	i. Survey ii. Interview iii. Online studies	i. Methodological ii. Meta-Analysis iii. Evaluation Studies iv. Operational Research v. KAP Studies vi. Impact Analysis vii. Implementation Research viii. Translational Research

4.7 Qualitative Research Methods

Qualitative research has a wider application in social sciences, humanities and other related areas. It mostly focuses on human behaviour from the point of view of participants. Qualitative research relates to the collection and analysis of non-numerical information/data like verbal or documented information to study concepts, perceptions, experiences, etc. of people on any event or topic of research. It has wide application in areas like anthropology, sociology, education, history and so on. The aim of qualitative research is to bring out the feeling, experience, insight and perception of people to throw light on any topic of study in this field. Qualitative research is a type of scientific research. In general terms, scientific research consists of an investigation that aims to seek the answer to a question systematically by using a pre-defined set of procedures to answer the question, collect evidence and produce findings that were not determined in advance. It produces findings that are applicable beyond the immediate boundaries of the study.

The strength of qualitative research is its ability to provide complex textual descriptions of how people experience a given research issue. It provides information about the 'human' side of an issue which is often contradictory to behaviours, beliefs, opinions, emotions and relationships of individuals. Qualitative methods of research are also effective in identifying intangible factors, such as social norms, socio-economic status, gender roles, ethnicity, religion, etc. whose role in the research issue may not be readily visible. When used along with quantitative methods, qualitative research can help the researcher interpret and better understand the complex reality of a given situation and the implications of quantitative data. The most commonly used qualitative methods are:

- Participant observation: It is appropriate for collecting data on naturally occurring behaviours in their usual contexts.
- In-depth interviews: It is optimal for collecting data on individuals' personal histories, perspectives and experiences, particularly when sensitive topics are being explored.
- Focus groups: These are effective in eliciting data on the cultural norms of a group and in generating broad overviews of issues of concern to the cultural groups or subgroups represented.

Each of these methods is particularly suited for obtaining a specific type of data.

As shown in Table 4.2, the quantitative and qualitative research methods differ primarily in their analytical objectives, the types of questions they pose, the types of data collection instruments they use, the forms of data they produce and the degree of flexibility built into study design.

One advantage of qualitative methods in exploratory research is that the use of open-ended questions and probing give participants the opportunity to respond in their own words, rather than forcing them to choose from given fixed responses/options. Open-ended questions have the ability to evoke responses that are:

- Meaningful and culturally salient to the participant.
- Unanticipated by the researcher.
- Explanatory in nature.

Another advantage of qualitative methods is that they allow the researcher the flexibility to probe initial participant responses—that is, to ask why or how. The researcher must listen carefully to what participants say, engage with them according to their individual personalities and

Table 4.2 Comparison of Quantitative and Qualitative Research Approaches

Particulars	*Quantitative*	*Qualitative*
Research hypothesis	Hypothesis formation possible in the form of numerical parameters.	Seek to explore a phenomenon under study.
Method of data collection	Use highly structured methods such as questionnaires, surveys and structured observation.	Use semi-structured methods such as in-depth interviews, focus groups and participant observation.
Data type	Research data in the form of countable or measurable values for dependent and independent variables. Numerical data (obtained by assigning numerical values to responses).	Participant's responses are collected through flexible methods and iterative style. Textual/verbal (from audiotapes, videotapes, field notes, etc.).
Questionnaire format	Generally closed-ended.	Mostly open-ended.
Goals and objectives	To describe parametric values of a population. To quantify variations around mean. To predict causal relationships among variables of study.	To describe patterns and relationships. To describe individual experiences. To describe group norms.
Analytical approach	Objective and deductive.	Subjective and inductive.
Nature of results	Accurate, quantifiable/numerical, stable, unbiased, reliable, etc.	Textural or word forms such as perception, insight, experience, etc.
Sample size	Relatively large.	Relatively small.
Nature of research design	Study design is stable throughout the study, participant responses are straightforward and independent of who, when and how the responses are sought, study design is stable subject to situation, assumptions and conditions.	Study design is not pre-planned and stable, participant's responses are complex and need streamlining, mixture of designs can be used to extract information.

styles, and use 'probes' to encourage them to elaborate on their answers. The following features of qualitative research are noteworthy.

- Qualitative research aims to study social and cultural phenomenon.
- It is an inductive approach to develop new concepts.
- Quantitative research designs emerge during the course of studies and not in advance.
- These are flexible and elastic.
- It typically leads to evolving methods of data collection.
- It requires more involvement of the researcher.

4.7.1 Types of Qualitative Observations

There are broadly four ways to collect qualitative observations:

- Complete observer—The researcher is completely unknown to the subjects of the study/focus group and even may not see each other.
- Observer as participant—Researcher is known to subjects of the study/focus group.
- Participant as observer—Complete indulgence of observer.
- Complete participant—Participants do not know the researcher.

4.7.2 Phases of Qualitative Research

The following are the phases of qualitative research:

- Orientation and review phase.
- Focused exploration.
- Confirmation and closure.

4.7.3 Data Collection Methods in Qualitative Research

The data collection methods in qualitative research include:

- Written expression by the participants.
- Observation by the researcher.
- Interactive interviews with the participants by the researcher.

4.7.4 Designs/Types of Qualitative Research

- Phenomenological research (how human awareness leads to social action—sociological).
- Ethnographic research (concentrates on the culture of a group of people -anthropological).
- Grounded theory (theories developed are grounded with information collected—any area).
- Historical models (causes, effects and trends related to past events to compare present events—historical).
- Narrative research (explore and conceptualize human experience in textual form—any area).
- Case study method (in-depth and detailed study of a social unit—a person, a family, an organization, a cultural group or a phenomenon in areas like social, educational, clinical, business, etc.).

More details about qualitative research are given in Chapter 9.

4.8 Mixed Methods

There is no water-tight demarcation for quantitative and qualitative research methods and designs. In some cases, both quantitative and qualitative methods are used by the researchers to cover all the objectives of the study. Besides some of the designs in survey and interview methods are used in both quantitative and qualitative research approaches.

4.9 Secondary Data Based Research

Periodical data are regularly collected by various Departments of State and Union Governments. The statistical abstracts of state and central governments are published every year which cover district-wise and state-wise data of various departments. Additionally, many ministries/departments like agriculture, energy, environment, education, finance, health and others have their own data-based periodical publications. The websites of most of the departments post the latest data on their sites. Many official and non-official organizations have their unpublished data in registers and other sources. All these can be important sources of data and can be sources of data for good research.

4.10 Multi-Disciplinary Research

Most of the research in applied areas is multi-disciplinary in nature. The research in agriculture sciences, medical sciences, business, etc. is multi-disciplinary in nature. While the research aims at cause-and-effect relationships, one has to go for support from other disciplines. Most of the institutional-level studies are carried out by a multi-disciplinary team of researchers.

The factors of morbidity patterns in medical sciences, factors for low productivity in agriculture, or factors for the volume of sale of a product, etc. can be better assessed through a multi-disciplinary research approach.

For most of the research problems, it may be possible to categorize under a specific method followed by the research design under that method. But in some research problems, a combination of different research methods and designs may be most appropriate. In some other cases, more than one research method or design may be required.

The research designs under different research methods, study purpose of each of these designs and features are given in Table 4.3.

More details about experimental research designs generally applied in health sciences are given in Chapter 5 and those in agricultural sciences are given in Chapter 6. The details of observational research designs are given in Chapter 7. Different types of surveys, interviews and online research designs are given in Chapter 8. The details of qualitative research designs are given in Chapter 9.

Table 4.3 Research Methods, Research Designs, Study Purposes, Situations and Features at a Glance

Research Method	*Research Design*	*Study Purpose*	*Situation/Conditions/ Features, etc.*
Experimental Method (applicable in health, agriculture, veterinary science education, etc.)	True experiment/ randomized control trial	To study effectiveness of one or more new treatment(s) compared to control/ standard one.	Sufficient number of homogenous experimental units (EUs) must be available. EUs are randomized as experimental and control groups.
	Post-test only	To study the outcome of a new treatment over control.	EUs are randomized as experimental and control groups, no pre-testing, treatment applied on experimental group only, post-testing of both groups. No baseline data generated.
	Pre-test post-test	To study effect of a new treatment over control.	Pretest observations from both groups, treatment applied on experimental group only, post-testing of both groups, baseline data available for comparison.
	Solomon four-group	To study the difference in effect of a new treatment over control.	Two experimental groups (E1 and E2), two control groups (C1 and C2), four groups randomly formed, pre-test observation from E1 and C1, treatment to E1 and E2, post-test observation from E1, E2, C1 and C2. More number of study units needed.
	Completely randomized design (CRD)	To study effectiveness of more than two new treatment(s) compared to standard based on homogenous units.	Homogenous EUs randomized as experimental and control groups, number of groups = number of treatments including control, number of units in each group need not be same.

(*Continued*)

Table 4.3 (Continued)

Research Method	*Research Design*	*Study Purpose*	*Situation/Conditions/ Features, etc.*
	Randomized block design (RBD)	To study effectiveness of more than two new treatment(s) compared to standard using heterogeneous units.	Within block homogenous EUs formed as sufficient such EUs not available for CRD, number of units in each block is equal to or multiple of the number of treatments.
	Latin square design (LSD)	To study effectiveness of more than two new treatment(s) compared to standard.	EUs arranged as rows and columns based on two-way variability factors. Number of treatments, rows and columns are all equal.
	Factorial design	To study effectiveness of a new treatment(s) as combination of factors compared to a standard.	Combination of factors with different levels for each and each combination is considered as a treatment. Interaction effect of factors can be ascertained.
	Quasi experiment design (non-randomized control trial)	To study effectiveness of a new treatment compared to standard.	Either randomization or control absent/not possible due to obvious reasons.
	Non-randomized control group design	To assess the effect of manipulated independent variables (IV) on outcome or dependent variable (DV).	Experimental and control groups will be there, but groups are not randomly formed. DV observed in both groups before and after treatment.
	Time series design	To measure effect of treatment over a period of time.	Only experimental group, observation of DV taken repeatedly for same treatment many times.
	Pre-experiment design	To assess response of treatment in a pilot manner.	No control group, it is a one-group experiment design.
	One shot post-test design	To assess response of treatment.	No control group and only one experimental group.
	One shot pre-test post-test design	To study the change in DV outcome due to treatment.	No control group and only one experimental group.
Non- Interventional Observation Research Methods (clinical, para-clinical and epidemiological)	Descriptive case reports	In-depth study of unique cases.	Rare and unique cases are studied for cause, treatments and outcome.
	Descriptive case series	To study a group of cases with similar clinical features.	Diagnosis, treatments and outcome without control.
	Descriptive cross-sectional	To study the cause factors of an event/disease in a population at a time.	Outcome analysis made without control with possible cause factors at a point in time.

(*Continued*)

Table 4.3 (Continued)

Research Method	*Research Design*	*Study Purpose*	*Situation/Conditions/ Features, etc.*
	Descriptive longitudinal	To study the cause factors of an event/disease in a population over time.	Outcome analysis made without control with possible cause factors over time.
	Analytical cross-sectional	To make exploratory, correlation, comparative studies, etc.	Outcome analysis made with or without control.
	Observational descriptive studies	To study distribution of diseases, identify characteristics and to form etiological hypothesis.	Population specified, disease defined and study can be census/sample based.
	Analytical case control	To identify risk factors attributable to cases/ diseases.	Homogenous cases and comparable control must be available.
	Analytical prospective cohort	To assess impact of risk factors leading to cases/diseases.	Start with exposure to risk factors to assess number of cases over time in exposure and control groups.
	Analytical retrospective cohort	To assess the number of cases in the past due to exposure to a risk factor.	Start with exposure and control groups and assess those who had cases in the past.
	Analytical ambispective cohort	To assess risk factor effect on cases.	Combination of the previous two, past from records and future cases observed over time.
	Community trials	To study interventional effect on aspects like community education on nutrition, family planning, breastfeeding, oral hygiene.	Experimental type with half of community as experimental and other half as control.
	Uncontrolled natural trials	To study the effect of natural calamities/ assess the problem.	Applicable to conditions like earthquake, famine, flood, etc.
	Before and after comparison trial	To study impact of vaccination, use of helmet, use of seat belt, etc.	Only one study group serves as control for before intervention and as experimental group after intervention.
Clinical Observational Method (patient-based)	Cross-sectional descriptive	To study status of defined outcome or dependent variable in relation to independent variable.	Based on OPD/IPD patients' medical observations with or without control.
	Cross-sectional comparative	To compare two or more situations for a defined outcome/ dependent variable.	Based on OPD/IPD patients' medical observations for M/F, R/U, etc.

(*Continued*)

Table 4.3 (Continued)

Research Method	*Research Design*	*Study Purpose*	*Situation/Conditions/ Features, etc.*
	Cross-sectional correlation	To correlate effect of dependent variable on outcome variables.	Based on OPD/IPD patients' medical observations.
	Cross-sectional exploratory	To explore cause of an outcome variable of the study.	Based on OPD/IPD patients' medical observations.
Survey and Interview Method (social sciences, business studies, commerce, education, etc.)	Questionnaire	Descriptive, analytical, explorative studies in which respondents fill information by themselves.	Literate respondents capable of knowing, expressing and recording facts and figures, opinions by themselves.
	Schedule	Descriptive, analytical, explorative studies in which information is filled by researcher/ representative.	Illiterate respondents or item of information need pre-explanation.
	Online	Descriptive, analytical, explorative studies in which respondents' downloads format (Google Form) fill information by themselves and upload it.	Respondents must be good at internet/e-communication.
	Telephonic	Descriptive, analytical, explorative studies in which information is collected on phone by researcher/ representative.	Both researcher and respondents must have access to phone/ mobile.
Secondary Data Based (all areas)	Spatial/temporal pattern of parameters under study	To study temporal pattern like trends, growth rates of variable under study and their spatial variations.	Based on available data from various sources (official and non-official).
Qualitative Research Methods	Phenomenological research	To study how human awareness leads to social action.	Based on live experiences of participants as described by them, researchers investigate a phenomenon or event.
	Ethnographic research	To study the life process of people.	Researchers become part of the people who are being studied to know their culture by living with them.
	Grounded theory	To study social processes and social structures of people.	Develop theories inductively.

(*Continued*)

Table 4.3 (Continued)

Research Method	*Research Design*	*Study Purpose*	*Situation/Conditions/ Features, etc.*
	Action research	To make social changes.	Researchers and participants jointly link theory to practice.
	Narrative research	To understand how participants perceive and make sense of their experiences.	Researchers share stories, incidents, etc. with the participants to assess/ know their perceptions and experiences.
Case Studies	'No Theory First' type case study	For methodological work of a first-time observed case.	To develop new theory/concept.
	Multiple case studies	For methodological work of a repeated rare case.	To develop new theory/concept.
	Comparative extreme cases	To analyze and compare independent variables of highly successful and failed cases and to identify factors causing it.	Extreme success and failure cases must be available.
Other Research Methods (all areas)	These include meta-analysis, evaluation studies, impact studies (with and without before and after studies), action research, developmental research, operational research, translational research, methodological research, KAP studies, etc.		

Some Important Points

- Research methods are the ways of conducting research for generating data, observations and evidence as per stipulated goals and objectives and in support of results/findings and conclusions/recommendations after the study is completed.
- The broad research methods include experimental (interventional), observational (non-interventional), survey and interview and situation-based research methods.
- Experimental research methods generally used in health and agricultural sciences include RCT, CRD, RBD, LSD, QED and PED.
- Observational methods are generally non-interventional methods used in health sciences (clinical and non-clinical), psychology, agriculture, economics, sociology, etc.
- The observational studies in clinical and para-clinical areas include descriptive, case reports, case series, cross-sectional, longitudinal and analytical studies.
- The observational epidemiological studies include case-control, cohort (prospective and retrospective), field trials, community trials, uncontrolled natural trials, natural experiments, etc.
- The survey, interview and online research methods include cross-sectional and longitudinal surveys, questionnaires, schedules, Google forms, telephonic interviews and online designs like focus groups, text analysis, social network analysis etc.
- Other situation-specific research methods include methodological, meta-analysis, evaluation studies, operational research, KAP studies, implementation research, transitional research, etc.
- The qualitative research methods include phenomenological, ethnographic, ground theory, historical models, narrative research, case study methods, etc.

Suggested Readings

Bernard, H. R., *Research methods in anthropology*, 2nd ed., Sage Publications, London, 1995.
Denzin, N. K., and Y. S. Lincoln (eds.), *Handbook of qualitative research*, Sage Publications, London, 2000.
Nkwi, P., I. Nyamongo, and G. Ryan, *Field research into social issues: methodological guidelines*, UNESCO, Washington, DC, 2001.
Pope, C., and N. Mays, *Qualitative research in health care*, BMJ Books, London, 2000.
Rao, S., and J. Richard, *Introduction to biostatistics and research methods*, 4th ed., Prentice-Hall, New Delhi, 2006.
Rao Sunder, P. S. S., *Introduction to biostatistics and research methods*, 5th ed., Prentice-Hall, New Delhi, 2012.
Sharma, B. S., *Research methods in social sciences*, Sterling Publishers Private Ltd., New Delhi, 1983.

5 Experimental (Interventional) Research Designs

5.1 Introduction

The main features of experimental research include presence of the experimental group(s) for one or more interventions, presence of a control group (with placebo, standard intervention or no intervention), random formation of both experimental groups and control group from the same population and collection of observations from all the groups. It has got wide range of applications in applied research in medical and health sciences, education, psychology, etc. In order to fit into application in different areas of research and also under different experimental conditions there are many designs under experimental methods, and these are discussed in this chapter.

5.2 True Experimental Design/Randomized Control Trials (RCT)

True experimental designs also known as randomized control trials are a group of interventional studies mostly in areas like health, agriculture, psychology, education, etc. In the categorization of research methods, RCTs are treated sometimes separately from true experiment types in view of their wide application in different areas of study. However, the principles of true experimental studies and RCTs are the same and hence treated at par by many researchers. The RCT is the gold standard for applied interventional research in many areas. The patients in health studies, students in educational research and the identified subjects of study in psychological studies, etc. constitute the units of study which are allocated randomly into study groups or experimental groups to receive one of the several intended interventions and a control group with either no intervention or a standard practice or placebo for comparison of results. It is basically to assess the efficacy of a new treatment, a new drug, a therapeutic or surgical procedure, etc. The intervention can also be the application of a treatment, drug or withdrawal of a drug in the experimental group. The outcome variable of the experimental group is compared with that of the control after the intervention is made and required information is collected from both the groups. This design allows assessing the impact on the outcome (dependent) variable due to the interventional (independent) variable. The participants in all the groups continue to be studied from the base period to the final stage of the study and if any unit in the study, for any reason, leaves in between the same are recorded and the study is continued with the remaining units.

The RCT designs can be used in specific areas like pharmacological, laboratory trials, clinical trials and field trials to evaluate preventive measures such as vaccination effect on large samples in natural settings. The RCTs are also classified based on interventional groups, control groups, aim of study, nature of research hypothesis, level of blindness to reduce the bias in the outcome of the trial and placebo effects.

DOI: 10.4324/9781003527183-5

In clinical trials, there can be three sources of bias leading to erroneous inference. (i) If the patients are aware that they are subjects of a new intervention, they may have the tendency to report improvement (ii) the observer also can have a similar attitude (iii) the statistician also can have a tendency to report positively. In order to overcome these biases 'Blinding Procedure' is adopted in experimental research. In 'single-blind trials' patients are not aware of the treatment; in 'double-blind trials' the patients and researcher/study team are not aware of the treatments given and in 'triple-blind trials' the patients, the researcher and the data analyst are blind about the identity of patients receiving the specific treatment. The knowledge of receiving/giving a new treatment can influence the respondent/researcher to distort the factual situation while recording information on dependent/outcome variables. This effect is called the placebo effect. Hence, the placebo intervention is made to overcome the placebo effect in research experiments.

In true experimental research/RCT, the researcher will have control over extraneous factors so as to assess with confidence the effect of treatment/intervention on outcome (dependent variable). The main features of the true experimental research design include:

- Presence of at least one intervention/treatment as manipulated by the researcher for application on subjects of study (treatment group) to observe its effect on outcome/dependent variable.
- Researcher will have control over extraneous factors.
- Presence of a control group without intervention but with any of three options—(i) negative control (no placebo) (ii) clear control (having placebo) (iii) positive control (with standard treatment). Except for the intervention, the control group units and experimental units will have perfect matching of other parameters. The statistical homogeneity test of the two groups for other parameters like anthropometric parameters will ensure matching of the two groups and also ensure the difference in outcome variable of the two groups as attributable to the intervention.
- Randomization in the allocation of subjects to a treatment group and control group is made to ensure that every subject has an equal chance of being included in the experimental or control groups. It helps to avoid systematic bias in the results of the experiment. It also helps to enhance both internal and external validity of the results of the research. Randomization can be done (i) by tossing a coin for each subject/plot of study, assuming outcome as head or tail deciding the subjects falling under treatment group or control group or vice versa (ii) by chit method of writing names or serial numbers of all subjects/plots in separate slips and drawing chits one after the other to place them in the experimental or control group (iii) by random number table or computer-generated random numbers after assigning serial numbers to the original group of combined units. Randomization can be simple (without restricting the number in each group), block randomization (restricting the number in each block of intervention to at least two times the number of treatments) or stratified randomization (to balance prognostic factors like age, sex, etc.). Generally, the same number of participants in the interventional and control groups is considered.

The various types of other true experiments depending upon the situation are:

5.2.1 ***Post-Test Only Control Design***

- Experimental and Control Groups are formed randomly from the homogenous group of subjects of study. The treatment or intervention is applied to the experimental group.
- Example: If the effect of special coaching a class is the problem under study, then two groups of students are formed randomly from among the students of the class. Those students who

attended special coaching classes form the experimental group and those who do not attend the special coaching classes form the control group. The same test is held for both the groups and the marks are obtained from the research data for further analysis.

- No pre-test is done before intervention.
- Post-test observations are recorded for both experimental and control groups and tested for significance.
- Difference in mean values of experimental and control groups tested using appropriate statistical tests. The layout of the post-test-only control design is shown in Figure 5.1.

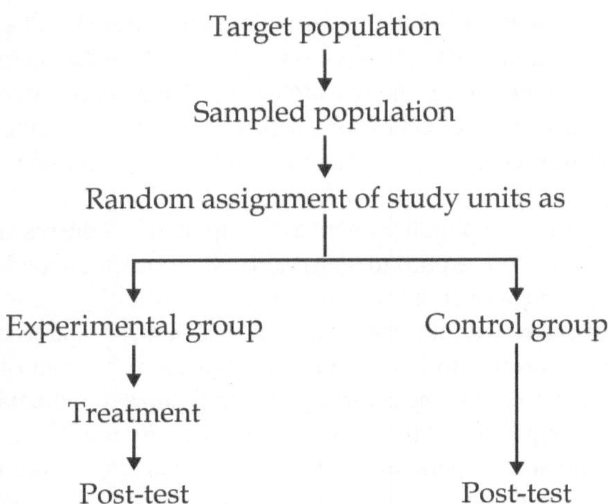

Figure 5.1 Diagram of Post-Test-Only Control Design

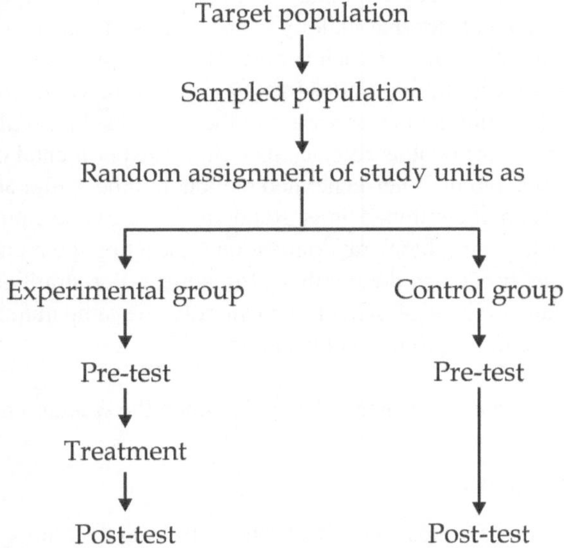

Figure 5.2 Diagram of Pre-Test Post-Test Design

5.2.2 Pre-Test Post-Test Design

- The intended number of subjects is randomly grouped into experimental and control groups.
- All units of both the groups are assessed before intervention is made on the experimental group which helps to confirm post-treatment effects with more confidence.
- Treatments (special coaching for the previous example) applied to subjects in the experimental group only.
- After giving special coaching to the treatment group both the groups are assessed for post-test performance.
- The post-test mean values are calculated and tested for significance.
- Examples: To assess the effect of a new teaching method, effect of counselling, effect of a drug to control an ailment. The layout of pre-test post-test design is given in Figure 5.2.

5.2.3 Solomon Four-Group Design

- It is a method developed by Richard Solomon in 1949. It is applicable in social sciences, psychology, medical sciences, etc. It is used to avoid the sensitization effect of pre-tests on subjects of study.

The four groups are:

Group 1: Pre-test, treatment and post-test.
Group 2: Pre-test, no treatment and post-test.
Group 3: No pre-test, treatment and post-test.
Group 4: No pre-test, no treatment and post-test only.

- Subjects are randomly assigned to these four groups.
- Group 1 and Group 2 are given pre-test.
- Group 1 and Group 3 are given treatment.
- Post-test observations are taken from all subjects of all four groups to assess the effect of treatment.
- The effectiveness of treatment can be assessed by comparing Group 1 with Group 3 as well as comparing Group 2 with Group 4.
- The two control groups are to reduce the influence of pre-test itself on the subjects.
- The Solomon four-group design is the combination of pre-test post-test design and the post-test only design.
- It guards against threats to possible internal and external validity.

Due to the difficulty in the formation of four random groups of more or less homogenous units, this design is not frequently used by researchers. The layout of this design is shown in Figure 5.3.

5.3 Quasi-Experimental Research (QER)/Non-Randomized Control Trials

Quasi experiment is used to assess the effect of manipulated independent variable on dependent or outcome variable. It is different from true experiments due to the absence of either randomization or a control group. It is used in such cases where randomization in the formation of experimental and control groups is not possible due to situational conditions, or control groups cannot be formed, or the complete controlled situation is not possible for manipulated independent variables.

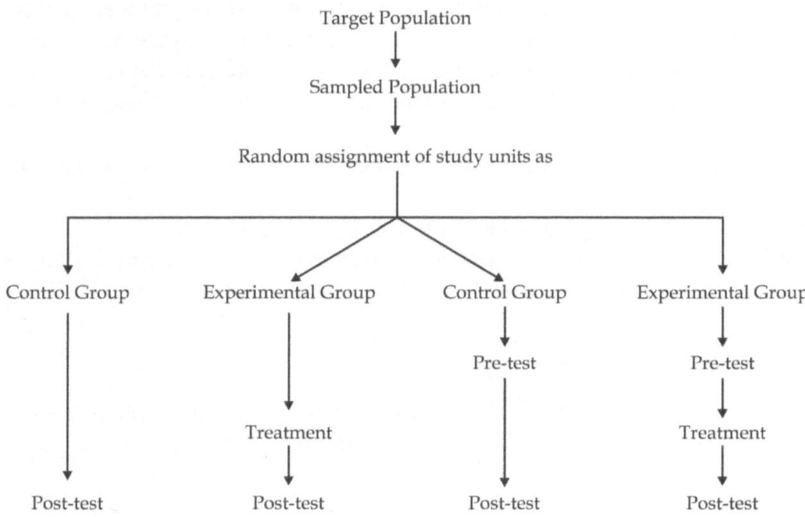

Figure 5.3 Diagram of Solomon Four-Group Design

The types of QER Designs are as under:

Non-Randomized Control Group Design/Non-equivalent Control Group Design:

- Experimental and control groups are formed without randomization.
- Looks like pre-test post-test true experiment design, but subjects are not randomly allocated to the groups.
- Dependent variable observed in experimental and control groups before intervention (pre-test).
- Subjects in the experimental group receive treatment.
- Post-test observations of the dependent variable are carried out for subjects in both groups to assess the effect of treatment/intervention.
- The layout of non-randomized control group quasi-experimental design is shown in Figure 5.4.

5.3.1 Time Series Design

It is used in such cases where treatment, over a period, has to be made by the researcher. It is possible to have this design on a single subject or a small group of similar subjects.

- When treatment effect is observed over a long period of time.
- Effects are recorded multiple times on a specific subject or on small groups during the experiment.
- Treatments are repeatedly given and withdrawn for a specified duration to assess their effectiveness.
- Example: Checking promptness of home assignments of students on a monthly basis.
- The layout of the time series design is given in Figure 5.5.

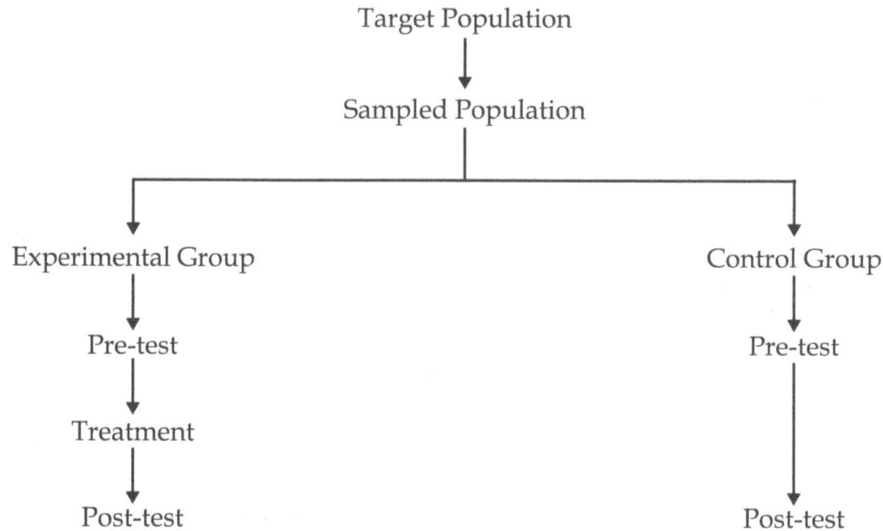

Figure 5.4 Diagram of Non-Randomized Control Group QER Design

Figure 5.5 Diagram of Time Series Quasi-Experimental Design

5.4 Pre-Experimental Research Design

Pre-experimental designs are not very powerful designs in research. It is like a pilot study in survey research for beginners in experimental research. It gives some idea of the proposed intervention. The two designs in vogue under this group of experimental designs are as under:

One-shot case design.

- It has only one experimental group to which treatment is applied, and post-test observations are taken to assess impact. No control group and random assignments of subjects.
- Normally held before conducting a true experiment as a pilot study for a true experiment to assess likely interventional response keeping in view the investment of cost and time for conduct of true experiment.
- It is a preliminary step to justify the proposed intervention and to ensure the effectiveness or potential of a true experiment.
- Example: Impact of a soft skill training to employees.
- The layout of the pre-experimental one-shot design is given in Figure 5.6.

One Group Pre-test Post-test Design:

- It has only one experimental group of study subjects. Observations are taken before and after treatment. Normally the subjects are randomly selected from the population to form the study group.

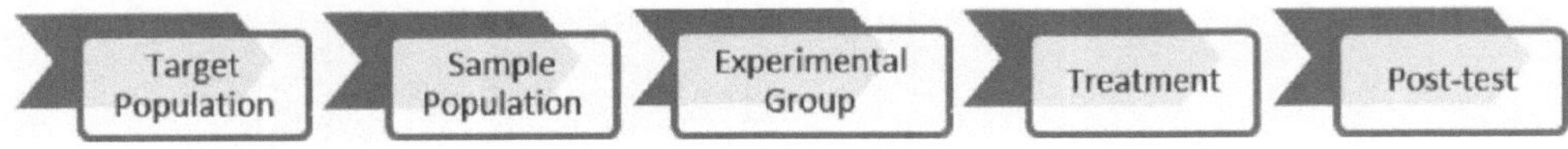

Figure 5.6 Diagram of One-Shot Pre-Experimental Design

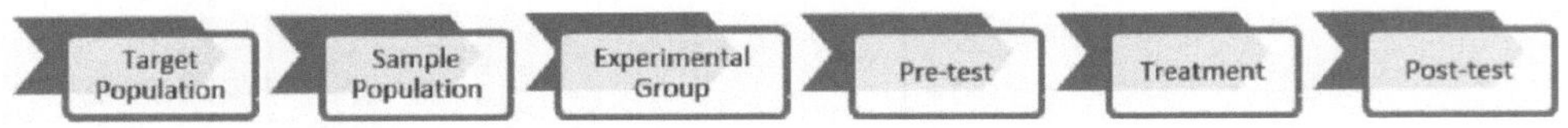

Figure 5.7 Diagram of One-Shot Pre-Test Post-Test Design

- Same type of assessment measures is taken before and after the treatment or exposed to a situation so as to assess the changes, if any, attributable to the treatment or condition.
- It is similar to quasi-experiments where instead of control and treatment group observations, before and after observations of the same experimental group are taken.
- Example: Effect of a new teaching method upon a group of children or impact of counselling.
- The layout of the design is given in Figure 5.7.

Some Important Points

- Experimental research designs are applied in areas like agriculture, health, education, psychology and similar areas to assess the impact of one or more interventions (as treatment/manipulation/addition/alteration) on the outcome variable of the study.
- The interventions can be a new practice, a new drug, a new combination of drugs, a new dose of a drug, or withdrawal of any drug or combinations, a new method of administering the inputs or drugs, a new method or process of treatment or surgery, a training, counselling and so on.
- Randomly formed experimental group(s) with intervention and control group without intervention from the same population are necessary.
- To avoid any form of bias by the researcher or subjects of study, adoption of phased blinding technique can be adopted.
- Different experimental designs for use in research in areas of medical sciences, education, psychology, etc. are available and in other areas per situational requirements.
- In true experimental research, the researcher will have full control over extraneous factors to assess with confidence the observed effect of treatment/intervention on outcome (dependent variable).
- **Randomized Control Trials (RCT)** have wide application in many areas, especially in clinical, epidemiology, agriculture, etc.
- **Post-Test Only Control Design:** Experimental and control groups are randomly formed, no pre-test is done before intervention, treatment/intervention is applied to the experimental group and post-test observations are recorded for both groups for analysis.
- **Pre-Test Post-Test Design:** The intended number of subjects is randomly grouped as experimental and control groups, pre-treatment observations are recorded for each subject in both groups, treatments are applied to subjects in the experimental group and after-treatment observations are recorded from subjects in both groups to assess the impact of treatment.

- **Solomon Four-Group Design:** Two experimental (E1 and E2) and two control (C1 and C2) groups are formed and subjects are randomly assigned to these four groups followed by treatments to all subjects of E1 and E2, post-test observations are taken from all subjects of all the four groups to assess the effect of treatment.
- **Quasi-Experimental Research (QER)/Non-Randomized Control Trials:** It differs from true experimental research with the absence of either randomization or a control group; even if a control group is present, random allocation of subjects may not be possible for obvious reasons.
- **Time Series Quasi-Experiment Design:** Is that in which treatment effect is observed over a long period of time.
- **Pre-Experimental Research Design:** It is like a one-shot case study design. It has only one experimental group on which treatment is applied and post-test observations are taken to assess impact.
- **One Group Pre-test Post-test Design:** It has only one experimental group of study subjects. Observations are taken before and after treatment. Normally the subjects are randomly selected from the population to form the group.

Suggested Readings

Campbell, D. T., and J. C. Stanley, 'Experimental and quasi-experimental designs for research on teaching', in *Handbook of research on teaching*, edited by N. L. Gage (ed.), Rand McNally, Chicago, IL, pp. 171–246, 1963.

Campbell, D. T., and J. C. Stanley, *Experimental and quasi-experimental designs for research*, Houghton Mifflin Company, Boston, MA, 1966.

Lachin, J. M., 'Statistical properties of randomization in clinical trials', *Controlled Clinical Trials* 9(4): 289–311, 1988.

Schulz, K. F., and O. A. Grimes, 'Generation of allocation sequences in randomized trials: chance, not choice', *Lancer* 359(9305): 515–519, 2002.

Sharma, B. S., *Research methods in social sciences*, Sterling Publishers Private Ltd., New Delhi, 1983.

Willmann, R., A. De Luca, M. Benatar, M. Grounds, J. Dubach, J.-M. Raymackers, and K. Nagaraju, 'Enhancing translation: guidelines for standard pre-clinical experiments in mdx mice', *Neuromuscular Disorders* 22: 43–49, 2012.

6 Design of Experiments for Field Research

6.1 Design of Experiments in Field Research—Important Concepts

Design of experiment is a research method to assess the outcome variable related to the treatments on the experimental units in such a way that the effect of each source of variation on the outcome variable can be estimated separately. It is the conduct of logical experiments under controlled situations to generate research data for drawing valid inferences/decisions. It is widely used to ascertain the response to interventions. The basic principles of this method include randomization, replication and local control. The experiments are conducted on already marked/identified units/plots. The intervention or object of assessment is termed as treatments which can be varieties of a crop, different doses of fertilizers, different levels of irrigation, different dates of sowing, different doses or types of plant protection chemicals, etc. in agricultural experiments; different procedures, drugs, doses of drugs, in relation to biological factors like age, sex etc. in health science; economic or social factors in social science. It is used mainly to select the best options/drugs/technologies out of all available alternatives/technologies.

Example: Effect of different drugs/products/practices/technologies to optimize production, early cure of a disease or to regulate the outcome variable of any clinical intervention.

The concepts of the following terms are important:

Null Hypothesis H_0: Effect of all treatments is the same.

Alternative Hypothesis H_1 (also known as research hypothesis): Effect of all treatments is not the same.

Experimental Material/Unit: Material/unit on which experiments are conducted, for example, patients/plants/animals/crop fields, etc.

Experimental Error: When the same treatment is applied to different units which are homogenous in nature, the values of the outcome variable may vary from unit to unit. This variation in outcome variable values due to uncontrolled factors is called experimental error. After extracting the variation due to all known factors remaining is known as experimental error. It is the basic principle behind the design of experiment analysis. It is always used to test the significance of differences. The estimated experimental error is used for the test of significance to assess the difference between the two treatments. If difference between two treatment means is greater than experimental error, then the difference is significant.

Randomization: Random allocation of treatments on experimental units for the validity of statistical tests or to avoid any bias in the experiment.

Replication: Repetition of the same treatment to estimate experimental error based on which final decisions are taken.

DOI: 10.4324/9781003527183-6

Local Control: Arrangement of experimental units into homogenous groups/blocks so that experimental units within the block are homogenous and whatever variation exists reflected in between the blocks (to compare treatment effects under the same situation). But, in the case of complete randomized block design (CRD), all the units should be homogenous as no block effect is there.

Precision of Design: It is the ability with which a design detects the small real difference between treatments. The lesser the experimental error, the greater is the precision of the design. The precision can be increased by increasing the number of replications and by selecting the most appropriate design for the experiment.

Accuracy of Design: It is the measure of lack of bias in the experimental design or the closeness of the estimated treatment effect to the true value of the population.

Degree of Freedom (DF): It is the difference between the number of observations used for the analysis and the number of independent constraints. For a design with '*k*' treatment, '*r*' replication and '*kr*' total observations '*kr−1*' is the DF for total; '*k−1*'is the DF for treatment '*r−1*' is the DF for replication and [(*kr−1*)−(*k−1*)−(*r−1*)] (or (*k−1*) × (*r−1*) if all treatments having equal number of replications and experimental design is RBD) is the DF for error. The sum of DF of different sources is equal to total DF.

Sum of Squares: It is the sum of the squared deviation of observations from the corresponding mean. Let X_1, X_2, X_3, ..., X_n are *n* observations with mean $\bar{X}$, then the sum of squares $= \sum (X - \bar{X})^2$.

F value: The F value is the ratio of variance generally; treatment variance is compared with error variance.

Analysis of Variance (ANOVA): When we have to make comparisons of more than two means, then ANOVA comes into the picture. It is the statistical technique used in the analysis of data under the design of an experiment where a number of hypotheses are tested together. The total variance in the set of observations is split into different components such as treatment, replication, error, etc. The research hypothesis for significant differences between the treatments and/or between the replications is tested together using error variance.

Test of Significance: The type of test of significance depends on the number of means of outcome variables to be compared. If we are having only one mean we cannot compare, but can test the validity of mean against a hypothetical value by applying the t-test. The significance of this test suggests that the mean is valid and can be used for prediction. We can also use the same test to compare the mean with a population mean 'μ'. Significance of this test indicates that the observed mean is significantly different from the population mean. If there are two means, say control and one treatment, we can apply an independent sample t-test. If observations are in pairs recorded on the same respondent/patient/unit at different time intervals, we apply a paired t-test. Paired t-test indicates the consistency of difference. However, if means are more than two, one can apply F-test only. The F and t are related statistics. In case of the two means, $F_{[1, d-1]} = t^2_{[d-1]}$ where 1 is the DF of the numerator and *d*−1 is the DF of the denominator.

Critical Difference (CD): If the null hypothesis is rejected by using ANOVA (*F*-test), it implies that all treatment effects are not the same. Then naturally the researcher would like to know which combinations of treatments differ significantly which can be known by **post hoc tests.** One of the post hoc test/criteria is critical difference (CD) the least significant difference (LSD). It is calculated by using standard error of the difference (SE_d) $\sqrt{\frac{2V_E}{r}}$. $CD = SE_d \times t_{[DF_e]}$.

The two treatment means having difference ≥CD, the difference is statistically significant.

Testing Procedure:

- As usual, the null hypothesis (H_0) and alternative hypothesis (H_1) are formed for all the sources of variation.
- Perform the test using 'F' statistic/analysis of variance (ANOVA). If 'F' is significant.
- Apply a post-hoc test for each source of variation separately to check the difference between any two means of a source.

Based on the number of treatments and the precision required, one may select the experimental design.
One treatment case: If there is only one treatment replicated *r* times, the applied test statistic is the t-test. In this case, we apply the treatment to a number of homogenous experimental units and observation records on these units. For five replication layout see Figure 6.1.

R1	R2	R3	R4	R5

Figure 6.1 Layout for One Treatment and Five Replication Case

In this case, all the five units are homogenous. No randomization is required as the same treatment is applied to the units.

In this experiment, we can test the validity of mean by applying the t-test $t_{[n-1]} = \frac{\bar{X}}{SE_m}$, significance of this t suggests that the mean is valid and can be used for prediction. We can also use this SE to compare this mean with the population mean μ $t_{[n-1]} = \frac{|\bar{X} - \mu|}{SE_m}$ where, $SE_m = \sqrt{\frac{MS}{n}}$; $MS = \frac{SS}{n-1}$; $SS = \sum_{i=1}^{n} X_i^2 - \frac{\left[\sum_{i=1}^{n} X_i\right]^2}{n}$. Significance of this t indicates that the observed mean is significantly different than the population mean.

Two treatments case: Two treatments $X^1$and X^2 are applied to n_1 and n_2 number of homogenous experimental units, respectively. The n_1 and n_2 may or may not be equal. Independent sample *t*-test is applicable to test the difference between two treatment means. If $X^1$and X^2 each applied to five replications, the layout can be seen in Figure 6.2.

X_1^1	X_3^1	X_5^1	X_2^1	X_4^1
X_5^2	X_3^2	X_1^2	X_4^2	X_2^2

Figure 6.2 Layout for Two Treatment and Five Replication Case

Here we apply an independent sample *t*-test $t_{[n_1+n_2-2]} = \frac{|\bar{X}^1 - \bar{X}^2|}{SE_d}$. Value of SE_d depends on the homogeneity of variance of $\bar{X}^1$ and $\bar{X}^2$. Calculate the mean square for both the groups as suggested and test the homogeneity of *MS* by *F*-test, that is, $F_{[DF_H, DF_L]} = \frac{MS_H}{MS_L}$; where DF_H and

DF_L are the degrees of freedom of higher and lower mean square (MS_H and MS_L) of mean 1 or 2. If F is non-significant MS is homogenous else heterogeneous. In the case of homogeneous MS the $SE_d = \sqrt{\frac{SS_1 + SS_2}{DF_1 + DF_2}}$ and compared with t at $DF_1 + DF_2$. In the case of heterogeneous $SE_d = \sqrt{MS_1 + MS_2}$ and compared with average t of t at DF_1 and t at DF_2. Significance of this can also be tested by the F-test. Significance of t and F suggest a significant difference between the two means.

Pair t-test: The treatment is applied to n homogenous experimental units. Observations are recorded prior to treatment and after the treatment. In such cases, treatment can be age, before and after procedure, training, etc. The layout for paired observations is shown in Figure 6.3.

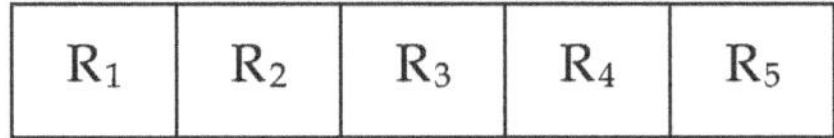

Figure 6.3 Layout for Paired Observation

Observations are recorded on each unit of study at T_1 and T_2 stages and apply paired t-test. Where the difference between each pair is calculated $D_i = X_i^1 - X_i^0$ and t-test is applied on this difference D_i. Calculate mean and SE for D_i values as suggested above and apply t-test i.e. $t_{[n-1]} = \frac{\bar{D}}{SE_m}$. Significance of this t suggests that the difference is consistent else varies from unit to unit (pair to pair).

Three or more treatments: If there are more than two means to be compared then one has to apply F-test. All the treatments are applied on the r uniform experimental units, or each treatment is replicated r times. The total experimental units are 'rt'. If all the rt units are homogenous, then we can apply CRD, where all the rt combinations (as each t treatment is replicated r times) are randomly applied on all the rt units. If not uniform, divide the units into r groups known as block or replication where all the treatments appear in each block. Again, the total number of units are rt. The design will be known as RBD. Even if r units are not uniform divide rt (r is replication and t is treatments) units in rc (r is rows and c is columns) groups. Here one treatment appears only once in a row and column. The experimental design is known as row-column design or lattice design. Accordingly, these three are the basic designs. Details of each design are explained:

6.2 Completely Randomized Design (CRD)

In CRD, the entire experimental area is homogenous, or all units are homogenous. CRD experiments are possible even for an unequal number of replications. Assuming an equal number of replications r for all t treatments, there are $t \times r$ units of experiments which are more or less homogenous. All the t treatments having r replications are randomly allocated to tr units at random that's why it is called a completely randomized design. The total variability in outcome variables is divided into sources such as treatment and error since the experimental units are homogeneous in CRD. In CRD any number of treatments can be used. Even if a few observations are missing due to any reason, the analysis of the remaining values can be done. In CRD the error degrees of freedom will be more as compared to other experimental designs. The major disadvantage of the CRD is the adherence to homogeneity of experimental units.

A	D	B	C	E
B	A	C	E	D
E	C	D	A	B

Figure 6.4 Layout for CRD with Five Treatments and Three Replications

Layout of experiment: Say there are five treatments (A, B, C, D and E) replicated three times. The layout is given in Figure 6.4. In this layout, the same treatment may appear in two adjoining units.

Preparation of data sheet: Data recorded on these treatments are arranged as shown in Table 6.1.

Table 6.1 Data Sheet for Replicated Treatments

Treatment	*R_1*	*R_2*	*Rr*	*Total*
1	X_{11}	X_{12}	X_{1r}	T_1
2	X_{21}	X_{22}	X_{2r}	T_2
...	...	...	...	...
t	X_{t1}	X_{t2}	X_{tr}	T_t

Steps of calculation: Calculation of different values for ANOVA table:

- $TDF = \left(\sum_{i=1}^{t} r_i\right) - 1$; If r is equal for all treatments, TDF = $rt - 1$.
- $DF_T = t - 1$
- $DF_E = TDF - DF_T$
- Mean for the ith treatment is $\bar{X}_i = \dfrac{\sum_{j=1}^{r} X_{ij}}{r_i}$
- Correction factor (CF): $CF = \dfrac{(X_{11} + X_{12} + \ldots + X_{tr})^2}{n}$; $n = r_1 + r_2 + \ldots + r_t$
- Or for an equal number of replications, the formula is $CF = \dfrac{\left(\sum_{i=1}^{t}\sum_{j=1}^{r} X_{ij}\right)^2}{tr}$ or $CF = \dfrac{GT^2}{tr}$
- Total sum of square (TSS) $= \left[\sum_{i=1}^{t}\sum_{j=1}^{r} X_{ij}^2\right] - CF$
- Treatment sum of square (SS_T) $= \left[\sum_{i=1}^{t} \dfrac{\left(\sum_{j=1}^{r} X_{ij}\right)^2}{r_i}\right] - CF$ for unequal number of replications
- OR $SS_T = \dfrac{\sum_{i=1}^{t}\left(\sum_{j=1}^{r} X_{ij}\right)^2}{r} - CF$ or $\dfrac{T_1^2 + T_2^2 + \ldots + T_k^2}{r} - CF$ for equal number of replications
- Error sum of square (SS_E) = $TSS - SS_T$

- Treatment mean square (MS_T) $= \frac{SS_T}{DF_T}$
- Error mean square (MS_E) $= \frac{SS_E}{DF_E}$
- F calculated $F_{[DF_T,\, DF_E]} = \frac{MS_T}{MS_E}$
- Calculate P with the help of Excel. Type "=FDIST(F,DF_T,DF_E)"

Put up all the values in the ANOVA table. The format is given in Table 6.2.

Table 6.2 The ANOVA Format for CRD

Source	*Degrees of freedom (DF)*	*Sum of squares (SS)*	*Mean squares (MS)*	*F*	*P*
Treatment	t–1	SS_T	MS_T	F_T	P_T
Error	n–t	SS_E	MS_E		
Total	n–1	TSS			

If $P < 0.05$ the H_0 is rejected and H_1 is accepted i.e. the treatment effect is significant and treatments are falling at least in two groups. To identify the difference between any two treatment means, apply the post-hoc test. Most common post-hoc test is the LSD test or CD.

The common critical difference (CD) for different levels of significance is CD 5% $= SEd \times t_{DF_E}$ 5% and CD 1% $= SEd \times t_{DF_E}$ 1%.

Where,

$$SE_d = \sqrt{\frac{MS_E}{r_i} + \frac{MS_E}{r_j}} \text{ for unequal number of replications}$$

$$SE_d = \sqrt{\frac{2MS_E}{r}} \text{ for equal number of replications}$$

t_{DFE} 5% and 1% are the table vale of t at error degrees of freedom and $\alpha = 0.05$ and 0.01, respectively.

Generally, CD 5% is used. If more precision is required, it can be at 1% or even less. The two treatment means having a difference greater than or equal to CD differs significantly. On the basis of significance, alphabets may be assigned to different treatment means for easy understanding. Two treatment means having non-significant differences assigned the same alphabet and if difference is significant, assign different alphabets.

Some important points of CRD: The CRD can be used for nutritional experiments like the effect of different products on weight gain by children if an adequate number of children with the same age and initial body weight are available. Some of the important points about CRD are:

- All experimental units' *n* are homogenous.
- There are a fixed number of treatments *t* for comparison.
- When the number of replications *r* for each treatment is the same, then $r = n/t$ (number of replications can be different also).
- Each treatment is applied randomly to the experimental units.

- Analysis under missing observation is possible.
- Provides maximum DF for error.
- Effect of different doses of medicine for recovery from the same illness of children.
- Effect of different drugs for recovery of patients of the same diseases.
- H_0: There is no significant difference between the treatment effects.
- H_1: Treatment effects are significantly different.

6.3 Randomized Block Design (RBD)

It may be difficult in most cases to get homogenous experimental units to conduct experiment in CRD. In RBD, experimental units are divided into blocks with homogenous units within the block. Each block is used as a replication and all the treatments will appear in each block. The total variability is divided into that of treatment, block/replication and error. As replications sum of square is extracted from the error the error is estimated with more precision in comparison to CRD.

RBD design can be used only when experimental units can be divided into homogenous groups/blocks. In agriculture, the research field is divided into different blocks each block having *t* plots. In social and health science blocks may be formed by putting same-age respondents/patients in one group. It allows any number of treatments and replications till homogenous units are available. The number of blocks is equal to the number of replications and the number of experimental units in each block is equal to the number of treatments.

Layout of RBD: For five treatments and three replications, the layout is given in Figure 6.5. Treatments are randomly allocated to experimental units in each replication.

R_1	A	D	B	C	E
R_2	B	A	C	E	D
R_3	E	C	D	A	B

Figure 6.5 Layout for RBD With Five Treatments and Three Replication Case

Preparation of data sheet: Data recorded on these treatments are arranged as shown in Table 6.3.

Table 6.3 Data Sheet for RBD

Treatment	R_1	R_2	Rr	*Total*
1	X_{11}	X_{12}	X_{1r}	T_1
2	X_{21}	X_{22}	X_{2r}	T_2
	...	...	...	
t	X_{t1}	X_{t2}	X_{tr}	T_t
Total	B_1	B_2	B_r	GT

Steps of calculation: Calculation of different values for ANOVA table:

- $TDF = rt - 1$
- $DF_R = r - 1$
- $DF_T = t - 1$

- $DF_E = TDF - DF_T - DF_R$ or $(t-1)(r-1)$
- Mean for the ith treatment is $\bar{X}_i = \dfrac{\sum_{j=1}^{r} X_{ij}}{r}$
- Correction factor (*CF*): $CF = \dfrac{(X_{11} + X_{12} + \ldots + X_{tr})^2}{tr}$; or $CF = \dfrac{\left(\sum_{i=1}^{t}\sum_{j=1}^{r} X_{ij}\right)^2}{tr}$ or $CF = \dfrac{GT^2}{tr}$
- Total sum of square (TSS) = $\left[\sum_{i=1}^{t}\sum_{j=1}^{r} X_{ij}^2\right] - CF$
- Replication sum of square (SS_R) = $\dfrac{\sum_{j=1}^{r}\left(\sum_{i=1}^{t} X_{ij}\right)^2}{t} - CF$ or $\dfrac{B_1^2 + B_2^2 + \ldots + B_r^2}{t} - CF$
- Treatment sum of square (SS_T) = $\dfrac{\sum_{i=1}^{t}\left(\sum_{j=1}^{r} X_{ij}\right)^2}{r} - CF$ or $\dfrac{T_1^2 + T_2^2 + \ldots + T_k^2}{r} - CF$
- Error sum of square (SS_E) = $\mathrm{TSS} - \mathrm{SS_T} - \mathrm{SS_R}$
- Treatment mean square (MS_T) = $\dfrac{SS_T}{DF_T}$
- Replication mean square (MS_T) = $\dfrac{SS_R}{DF_R}$
- Error mean square (MS_E) = $\dfrac{SS_E}{DF_E}$
- F calculated for replications (F_R) $F_{[DF_R,\, DF_E]} = \dfrac{MS_R}{MS_E}$
- F calculated for treatments (F_T) $F_{[DF_T,\, DF_E]} = \dfrac{MS_T}{MS_E}$
- Calculate P for replications with the help of Excel. Type "=FDIST(F,DF_R,DF_E)"
- Calculate P for treatments with the help of Excel. Type "=FDIST(F,DF_T,DF_E)"

Put up all the values in the ANOVA table as shown in Table 6.4.

Table 6.4 The ANOVA Format for RBD

Source	*Degrees of freedom (DF)*	*Sum of squares (SS)*	*Mean squares (MS)*	*F*	*P*
Replication	*r*–1	SS_R	MS_R	F_R	P_R
Treatment	*t*–1	SS_T	MS_T	F_T	P_T
Error	(*r*–1)(*t*–1)	SS_E	MS_E		
Total	*rt*–1	TSS			

If P_R and/or $P_T < 0.05$, the H_0 is rejected and H_1 is accepted for replications and/or treatments i.e. replication and/or treatment effect(s) is/are significant. To identify the difference between any two treatments mean apply the post-hoc test. Most common post-hoc test is the LSD test.

LSD or CD 5% $= SEd \times t_{DF_E}$ 5% and CD 1% $= SEd \times t_{DF_E}$ 1%

Where, $SE_d = \sqrt{\dfrac{2MS_E}{r}}$

Generally, CD 5% is used. If more precision is required, it can be at 1% or even less. The two treatment means having the difference greater than or equal to CD differs significantly. On the basis of significance, alphabets may be assigned to different treatment means for easy understanding. Two treatment means having non-significant differences are assigned same alphabet else different.

Some important points of RBD: The RBD design can be used for any experiment where experimental units are divided into *r* homogenous blocks. Some of the important points about RBD are:

- All experiment units are not homogenous.
- They are grouped into blocks.
- All the units in a block are homogenous.
- Each block is one replication.
- Treatments are applied randomly to units within each block.
- Total variability is split into three i.e. replication, treatments and error.

Example:

(i) Effect of different drugs in controlling some illnesses of patients. Different wards may be treated as replication.

(ii) Impact of different baby foods for balanced growth of underweight children. Group of children from each selected location forms a replication.

- Two hypotheses tested in ANOVA i.e. one for replication and another for treatment:
    - H_0: No significant difference in the effect of drugs/baby food and no significant difference between replications.
    - H_1: Significant effect of drugs/baby food and significant difference between replications.

6.4 Latin Square Design (LSD)

LSD is another basic design where entire experimental units are divided into horizontal (rows) and vertical (columns) blocks having homogenous units in both. In nutritional trials, the children can be horizontally grouped according to weight and vertically grouped according to age. Here, each row (horizontal group) and each column (vertical group) is a replication. Hence, the number of rows = number of columns = number of treatments and hence it is called Latin Square Design or row-column design. Here the total variability is divided into rows (weight), columns (age) and treatments. LSD makes it possible to eliminate errors in two ways from experimental units and reduces experimental errors accordingly.

Layout of LSD: For five treatments, the layout is given in Figure 6.6. Treatments are randomly allocated to experimental units in rows and columns in such a way so that treatment appears only once in a row or in a column.

Preparation of data sheet: Data recorded on these are arranged in two two-way tables. One for treatments effects, Table 6.5 and another for rows and column effects, Table 6.6.

Column/ Row	C_1	C_2	C_3	C_4	C_5
R_1	A	D	B	C	E
R_2	B	A	C	E	D
R_3	E	C	D	A	B
R_4	D	E	A	B	C
R_5	C	B	E	D	A

Figure 6.6 Layout for LSD with Five Treatment Cases

Table 6.5 Data Sheet for LSD Treatment Effects

Treatment	R_1	R_2	*Ri*	*Rr*	*Total*
1 = A	X_{11}	X_{12}	X_{1i}	X_{1r}	T_1
2 = B	X_{21}	X_{22}	X_{2i}	X_{2r}	T_2
	. . .	. . .	. . .	. . .	
t = E	X_{t1}	X_{t2}	X_{ti}	X_{tr}	T_t

Table 6.6 Data Sheet for LSD Rows and Column Effects

	Columns					
Rows	C_1	C_2	C_3	C_4	R_5	*Row Total*
R_1	X_{11}	X_{12}	X_{13}	X_{14}	X_{15}	R_1
R_2	X_{21}	X_{22}	X_{23}	X_{24}	X_{25}	R_2
R_3	X_{31}	X_{32}	X_{33}	X_{34}	X_{35}	R_3
R_4	X_{41}	X_{42}	X_{43}	X_{44}	X_{45}	R_4
R_5	X_{51}	X_{52}	X_{53}	X_{54}	X_{55}	R_5
Column Total	C_1	C_2	C_3	C_4	C_5	GT

Steps of calculation: Calculation of different values for ANOVA table:

- $TDF = rc - 1$
- $DF_R = t - 1$
- $DF_C = t - 1$
- $DF_T = t - 1$
- $DF_E = TDF - DF_R - DF_C - DF_T$ or $(t-1)(t-2)$
- Mean for the ith treatment is $\bar{X}_i = \frac{\sum_{j=1}^{r} X_{ij}}{r}$ or $\bar{X}_i = \frac{T_1}{r}$ using Table 6.5

- Correction factor (*CF*) = $\frac{\left(\sum_{i=1}^{c}\sum_{j=1}^{r}X_{ij}\right)^2}{cr}$ or $\frac{GT^2}{cr}$ using Table 6.6
- Total sum of square (TSS) = $\sum_{i=1}^{c}\sum_{j=1}^{r}X_{ij}^2 - CF$ or $X_{11}^2 + X_{12}^2 + ... + X_{55}^2 - CF$ using Table 6.6
- Row sum of square (SS_R)= $\frac{\sum_{j=1}^{r}\left(\sum_{i=1}^{c}X_{ij}\right)^2}{t} - CF$ or $\frac{R_1^2 + R_2^2 + ... + R_r^2}{t} - CF$ using Table 6.6
- Column sum of square (SS_C)= $\frac{\sum_{i=1}^{c}\left(\sum_{j=1}^{r}X_{ij}\right)^2}{t} - CF$ or $\frac{C_1^2 + C_2^2 + ... + C_c^2}{t} - CF$ using Table 6.6
- Treatment sum of square (SS_T)= $\frac{\sum_{i=1}^{t}\left(\sum_{j=1}^{r}X_{ij}\right)^2}{r} - CF$ or $\frac{T_1^2 + T_2^2 + ... + T_t^2}{t} - CF$ using Table 6.5
- In present case, $r = c = t = 5$ and T_1 to T_5 are A, B, C, D and E treatment
- Error sum of square (SS_E) = $TSS - SS_T - SS_R - SS_C$
- Row mean square (MS_R) = $\frac{SS_R}{DF_R}$
- Column mean square (MS_C) = $\frac{SS_C}{DF_C}$
- Treatment mean square (MS_T) = $\frac{SS_T}{DF_T}$
- Error mean square (MS_E) = $\frac{SS_E}{DF_E}$
- F calculated for row (F_R) $F_{[DF_R,\ DF_E]} = \frac{MS_R}{MS_E}$
- F calculated for column (F_C) $F_{[DF_C,\ DF_E]} = \frac{MS_C}{MS_E}$
- F calculated for treatments (F_T) $F_{[DF_T,\ DF_E]} = \frac{MS_T}{MS_E}$
- Calculate P for rows with the help of Excel. Type "=FDIST(F_R,DF_R,DF_E)"
- Calculate P for columns with the help of Excel. Type "=FDIST(F_C,DF_C,DF_E)"
- Calculate P for treatments with the help of Excel. Type "=FDIST(F_T,DF_T,DF_E)"

Put up all the values in the ANOVA table as shown in Table 6.7.

Table 6.7 The ANOVA Format for LSD

Source	*Degrees of freedom (DF)*	*Sum of squares (SS)*	*Mean sum of squares (MSS)*	*F*	*P*
Rows	$t-1$	SS_R	MS_R	F_R	P_R
Columns	$t-1$	SS_C	MS_C	F_C	P_C
Treatment	$t-1$	SS_T	MS_T	F_T	P_T
Error	$(t-1)(t-2)$	SS_E	MS_E		
Total	t^2-1	TSS			

If P of any source < 0.05, the H_0 is rejected and H_1 is accepted, that is, the difference is significant. To identify the difference between any two treatment mean, apply the post-hoc test. Most common post-hoc test is the CD test.

$$\text{CD } 5\% = SE_d \times t_{DF_E} 5\% \text{ and CD } 1\% = SE_d \times t_{DF_E} 1\%$$

Where,

$$SE_d = \sqrt{\frac{2MS_E}{r}}$$

CD 5% is most common. If more precision is required, it can be at 1% or less. The two-treatment means having a difference greater than or equal to CD differs significantly. On the basis of significance, alphabets may be assigned to different treatment means for easy understanding. Two treatment means having non-significant differences are assigned the same alphabet else different.

Some important points of LSD: The LSD design can be used for a smaller number of treatments where experimental units are divided into rows and columns. In this way, the error estimated in this design is more precise than CRD and RBD. Some of the important points about LSD are:

- Experiments units are not homogenous.
- Units are divided horizontally as well as vertically in rows and columns.
- Each row and column having *t* units.
- Treatments are applied randomly on units in such a way so that treatment appear only once in a row and in a column.
- Total variability is split into four i.e. rows, columns, treatments and error.

Example:

(i) Effect of different drugs in controlling some illnesses of patients of different ages and weight groups.
(ii) Impact of different baby foods in balanced growth of underweight children in different age and body weight groups.

Here each row has *t* patients of equal age, and each column has *t* patients of equal weight.

- Error estimated is more precise than CRD and RBD.
- Here we test three hypotheses: one for treatments, second for rows and third for columns:
  - H_0: No significant difference in the effect of drugs/baby food, between rows and between columns.
  - H_1: Significant difference between drugs/baby food, between rows and between columns.

Apart from basic designs, designs are also classified on the basis of the number of sources of variation such as treatments, year, location, environments, etc. and partition the treatments accordingly. Analysis of designs depends on the nature of the source of variation. Say the experiment is repeated at different locations then the number of replications maasdgadsy be the same but the effect of replications at one location may be different than at another location because experimental units are different at both locations. But, treatments or combinations of treatments remain the same at all locations. So, both sources are analyzed in different ways. Generally, treatment SS is partitioned according to components of treatments, namely, the drugs and their doses, fertilizers and doses, etc., that is the effect of drugs, concentration and interaction are tested. In health science, for example, to treat a disease, four drugs (factor one) are used each with three uniform concentrations (factor second) and all patients are homogenous, that is, the basic design is CRD and this design is known as two-factor factorial CRD design. If these treatments are evaluated in RBD, it will be known as two-factor factorial RBD where patients are divided in *r* homogenous units and each treatment is applied in all units of each replication. If the concentrations of drugs are not the same, such as different drugs having different concentrations even if the number of concentrations are the same, the design is known as nested either two-factor nested CRD or RBD depending on the use of the basic design. In this way, we can have the number of factors and designs named accordingly.

6.5 Factorial Design

When all the members of one factor (drugs, D) are evaluated in combination with all members of another factor (concentration, C) is known as factorial design D × C. Here we can assess the effect of factors D and C separately and the interaction of D and C. These designs are required to assess variations in two or more factors simultaneously through the same experiment. When a combination of two drugs (A and B), each with varying doses a_1 and a_2 for A and b_1 and b_2 for B having combination a_1b_1, a_1b_2, a_2b_1 and a_2b_2 will be considered as separate treatments and the experiments are to be planned accordingly instead of conducting two separate experiments for drugs A and B. The factorial design has the advantage of getting the interaction effect as well as the main effect of the two factors. Factorial experiments are valid even if the factors are independent without any interaction effect.

Layout: Layout depends on the selection of basic design. As RBD is the most popular we select it. For four treatments combinations (A = a_1b_1, B = a_1b_2, C = a_2b_1 and $D = a_2b_2$) three replications layout is given in Figure 6.7. Treatment combinations are randomly allocated to experimental units in each replication.

Preparation of data sheet: Data recorded on these treatments are arranged as shown in Table 6.8.

To calculate the main effects and interactions, the total over replications are arranged in another table having *a* rows and *b* columns as shown in Table 6.9.

Steps of calculation: Calculation of different values for ANOVA table:

- $TDF = rt - 1$
- $DF_R = r - 1$
- $DF_T = ab - 1$
- $DF_A = a - 1$

R_1	A	D	B	C
R_2	B	A	C	D
R_3	C	D	A	B

Figure 6.7 Layout for Two-Factor Factorial Design with Three Replication Case

Table 6.8 Data Sheet for Factorial Design

Treatment	R_1	R_2	Rr	*Total*
a_1b_1	X_{111}	X_{112}	X_{11r}	$\sum_{k=1}^{r} X_{11k}$
a_1b_2	X_{121}	X_{122}	X_{12r}	$\sum_{k=1}^{r} X_{12k}$
a_2b_1	X_{211}	X_{212}	X_{21r}	$\sum_{k=1}^{r} X_{21k}$
a_2b_2	X_{221}	X_{222}	X_{22r}	$\sum_{k=1}^{r} X_{22k}$
Total	$\sum_{i=1}^{a}\sum_{j=1}^{b} X_{ij1}$	$\sum_{i=1}^{a}\sum_{j=1}^{b} X_{ij2}$	$\sum_{i=1}^{a}\sum_{j=1}^{b} X_{ijr}$	$\sum_{i=1}^{a}\sum_{j=1}^{b}\sum_{k=1}^{r} X_{ijk}$

Table 6.9 Two-Way Table for Two-Factor Two-Level Factorial Experiment

Treatment	b_1	b_2	*Total*
a_1	$\sum_{k=1}^{r} X_{11k}$	$\sum_{k=1}^{r} X_{12k}$	$\sum_{j=1}^{b}\sum_{k=1}^{r} X_{1jk}$
a_2	$\sum_{k=1}^{r} X_{21k}$	$\sum_{k=1}^{r} X_{22k}$	$\sum_{j=1}^{b}\sum_{k=1}^{r} X_{2jk}$
Total	$\sum_{i=1}^{a}\sum_{k=1}^{r} X_{i1k}$	$\sum_{i=1}^{a}\sum_{k=1}^{r} X_{i2k}$	$\sum_{i=1}^{a}\sum_{j=1}^{b}\sum_{k=1}^{r} X_{ijk}$

- $DF_B = b - 1$
- $DF_{AB} = (a-1)(b-1)$
- $DF_E = TDF - DF_T - DF_R$ or $(\text{ab}-1)(r-1)$
- Mean for interaction between i^{th} drug and j^{th} concentration $\bar{X}_{ij} = \frac{\sum_{k=1}^{r} X_{ijk}}{r}$
- Mean for i^{th} drug over b concentrations and r replications $\bar{A}_i = \frac{\sum_{j=1}^{b}\sum_{k=1}^{r} X_{ijk}}{br}$
- Mean for j^{th} concentration over a drugs and r replications $\bar{B}_j = \frac{\sum_{i=1}^{a}\sum_{k=1}^{r} X_{ijk}}{ar}$

- Overall experimental mean $\bar{X} = \dfrac{\sum_{i=1}^{a}\sum_{j=1}^{b}\sum_{k=1}^{r} X_{ijk}}{abr}$
- Correction factor (CF): $CF = \dfrac{\left(\sum_{i=1}^{a}\sum_{j=1}^{b}\sum_{k=1}^{r} X_{ijk}\right)^2}{abr}$
- Total sum of square (TSS): $TSS = \sum_{i=1}^{a}\sum_{j=1}^{b}\sum_{k=1}^{r} X_{ijk}^2 - CF$
- Replication sum of square (SS_R) $= \dfrac{\sum_{k=1}^{r}\left(\sum_{i=1}^{a}\sum_{j=1}^{b} X_{ijk}\right)^2}{ab} - CF$
- Treatment sum of square (SS_T) $= \dfrac{\sum_{i=1}^{a}\sum_{j=1}^{b}\left(\sum_{k=1}^{r} X_{ijk}\right)^2}{r} - CF$
- A sum of square (SS_A) $= \dfrac{\sum_{i=1}^{a}\left(\sum_{j=1}^{b}\sum_{k=1}^{r} X_{ijk}\right)^2}{br} - CF$
- B sum of square (SS_B) $= \dfrac{\sum_{j=1}^{b}\left(\sum_{i=1}^{a}\sum_{k=1}^{r} X_{ijk}\right)^2}{ar} - CF$
- A × B sum of square (SS_{AB}) $= SS_T - SS_A - SS_B$
- Error sum of square (SS_E) $= TSS - SS_T - SS_R$
- Mean square of ith source (MS_i) $= \dfrac{SS_i}{DF_i}$
- F calculated for i^{th} source (F_i) $F_{[DF_i,\, DF_E]} = \dfrac{MS_i}{MS_E}$
- Calculate P for i^{th} source with the help of Excel. Type "=FDIST (F_i, DF_i, DF_E)"

Put up all the values in the ANOVA table as shown in Table 6.10.

If P < 0.05 the H_0 is rejected and H_1 is accepted for any source, that is, replication, treatment, A, B and A × B effect is said to be significant. To identify the difference between any two

Table 6.10 The ANOVA Format for Two-Factor Factorial RBD

Source	*Degrees of freedom (DF)*	*Sum of Squares (SS)*	*Mean sum of squares (MSS)*	*F*	*P*
Replication	*r−1*	SS_R	MS_R	F_R	P_R
Treatment	*ab−1*	SS_T	MS_T	F_T	P_T
A	*a−1*	SS_A	MS_A	F_A	P_A
B	*b−1*	SS_B	MS_B	F_B	P_B
A × B	*(a−1)(b−1)*	SS_C	MS_C	F_C	P_C
Error	*(ab−1)(r−1)*	SS_E	MS_E		
Total	*abr*−1	TSS			

treatments mean apply the post-hoc test. Most common post-hoc test is the LSD test. Formula of SE_d is different for different sources.

SE_d for the difference between any two interactions means $SE_{d_{A\times B}} = \sqrt{\frac{2MS_E}{r}}$

SE_d for the difference between any two A means $SE_{d_A} = \sqrt{\frac{2MS_E}{br}}$

SE_d for the difference between any two B means $SE_{d_B} = \sqrt{\frac{2MS_E}{ar}}$

CD_i $5\% = SEd_i \times t_{DF_E} 5\%$ and CD $1\% = SEd_i \times t_{DF_E} 1\%$

CD 5% is most common. If more precision is required, it can be at 1% or less. The two-treatment means having a difference greater than or equal to CD differs significantly. On the basis of significance alphabets may be assigned to different treatment means for easy understanding. Two-treatment means having non-significant differences assigned same alphabet or different.

Some important points of factorial design: The two-factor factorial CRD/RBD design can be used for any experiment experimenter who is interested in knowing the effect of factor A, factor B and their interaction A × B. Some important points about factorial design are:

- All the members of factor A having equal number of B factors.
- Each member of factor A having combination with each member of factor B.
- Treatment SS is split in A, B and A × B SS.
- The ab treatment combinations are evaluated in any basic experimental design, CRD or RBD depending upon the requirement and experimental conditions.
- Apart from the hypothesis of basic designs, treatment in CRD and; replication and treatment in RBD, another hypothesis tested in ANOVA are:
    - H_0: No significant difference in levels of A, B and A ×B.
    - H_1: Significant difference in levels of A, B and A × B.
- Difference between the two means is tested with respective CD.
- If two means having a difference of more than or equal to CD, the difference is significant.
- Factors can be more than two. For three-factor treatment, SS is split as A, B, C, A × B, A × C, B × C and A × B × C.
- Levels of A, B and C may be different but remain the same for all members of other factors.
- Number of factors can be more than three.

6.6 Nested Design

When the members of factor B vary with levels of factor A in a factorial design it is called nested design, for example, in health science, dosages vary with drugs, in agriculture, levels vary with fertilizer and are evaluated in nested designs. In this design, sources are between A, between B within A_1, A_2, A_3, etc. The interaction effect is not there. Levels of B may or may not vary for different A's. For drug or fertilizer, A_1 levels of B are B_1, B_2, B_3, etc. and A_2 levels are $B_{1'}$, $B_{2'}$, $B_{3'}$, etc. In other words, dosages of drug A_1 are 1, 10 and 15 mg/kg body weight and for drug A_2 are 2, 8 and 12 mg/kg body weight. So, between concentrations, comparison is not possible across the drugs. Between concertation, comparison is possible only within each A's or possible only within the drug. Accordingly, interaction A × B is also not possible.

Layout: Layout of nested designs is the same as for factorial designs and depends on the selection of the basic design. For two drugs with two levels of concentrations, four treatment combinations are (A = a_1b_1, B = a_1b_2, C = a_2b_1, and D = a_2b_2) and a three replication layout is given in Figure 6.8. Treatment combinations are randomly allocated to experimental units in each replication.

R1	A	D	B	C
R2	B	A	C	D
R3	C	D	A	B

Figure 6.8 Layout for Two-Factor Nested Design with Three Replication Case

Preparation of data sheet: Data recorded are arranged in treatment by replication as shown in Table 6.11.

Table 6.11 Data Sheet for Two-Factor Two-Level Nested Design

Treatment	R_1	R_2	*Rr*	*Total*
a_1b_1	X_{11}	X_{12}	X_{1r}	$\sum_{j=1}^{r} X_{1j}$
a_1b_2	X_{21}	X_{22}	X_{2r}	$\sum_{j=1}^{r} X_{2j}$
a_2b_1	X_{31}	X_{32}	X_{3r}	$\sum_{j=1}^{r} X_{3j}$
a_2b_2	X_{41}	X_{42}	X_{4r}	$\sum_{j=1}^{r} X_{4j}$
Total	$\sum_{i=1}^{t} X_{i1}$	$\sum_{i=1}^{t} X_{i2}$	$\sum_{i=1}^{t} X_{ir}$	$\sum_{i=1}^{t}\sum_{j=1}^{r} X_{ij}$

To calculate the SS for different sources total over replications is arranged in Table 6.12 having *a* rows and *b* columns but the total for each level of B over A is not required.

Table 6.12 Two-Way Table for Two-Factor Two-Level Nested Experimental Design

Treatment	b_1	b_2	*Total*
a_1	$\sum_{k=1}^{r} X_{1k}$	$\sum_{k=1}^{r} X_{2k}$	$\sum_{j=1}^{b_1}\sum_{k=1}^{r} X_{1jk}$

(Continued)

Table 6.12 (Continued)

Treatment	b_1	b_2	*Total*
$\mathbf{a_2}$	$\sum_{k=1}^{r} X_{3k}$	$\sum_{k=1}^{r} X_{4k}$	$\sum_{j=1}^{b_2}\sum_{k=1}^{r} X_{2jk}$
Total			$\sum_{i=1}^{a}\sum_{j=1}^{b_i}\sum_{k=1}^{r} X_{ijk}$

Steps of calculation: Calculation of different values for ANOVA table:

- $TDF = rt - 1$
- $DF_R = r - 1$
- $DF_T = t - 1$
- $DF_A = a - 1$
- $DF_{B_i} = b_i - 1$
- $DF_E = TDF - DF_T - DF_R$ or $(t-1)(r-1)$
- Mean of concentration B_j within drug A_i or i^{th} treatment $\bar{X}_{ij} = \frac{\sum_{k=1}^{r} X_{ijk}}{r}$ or $\frac{\sum_{j=1}^{r} X_{ij}}{r}$
- Mean for i^{th} drug over b_i concentrations and r replications $\bar{A}_i = \frac{\sum_{j=1}^{b_i}\sum_{k=1}^{r} X_{ijk}}{rb_i}$
- Experimental mean overall $\bar{X} = \frac{\sum_{i=1}^{a}\sum_{j=1}^{b_i}\sum_{k=1}^{r} X_{ijk}}{r\sum_{i=1}^{a} b_i}$
- Correction factor (CF): $CF = \frac{\left(\sum_{i=1}^{a}\sum_{j=1}^{b_i}\sum_{k=1}^{r} X_{ijk}\right)^2}{r\sum_{i=1}^{a} b_i}$ or $\frac{\left(\sum_{i=1}^{t}\sum_{j=1}^{r} X_{ij}\right)^2}{tr}$
- Total sum of square (TSS): $TSS = \sum_{i=1}^{a}\sum_{j=1}^{b}\sum_{k=1}^{r} X_{ijk}^2 - CF$ or $\sum_{i=1}^{t}\sum_{j=1}^{r} X_{ij}^2 - CF$
- Replication sum of square (SS_R) $= \frac{\sum_{k=1}^{r}\left(\sum_{i=1}^{a}\sum_{j=1}^{b} X_{ijk}\right)^2}{\sum_{i=1}^{a} b_i} - CF$ or

$$\frac{\sum_{j=1}^{r}\left(\sum_{i=1}^{t} X_{ij}\right)^2}{t} - CF$$

- Treatment sum of square (SS_T) $= \frac{\sum_{i=1}^{a}\sum_{j=1}^{b_i}\left(\sum_{k=1}^{r} X_{ijk}\right)^2}{r} - CF$ or

$$\frac{\sum_{i=1}^{t}\left(\sum_{j=1}^{r} X_{ij}\right)^2}{r} - CF$$

- A sum of square (SS_A) $= \left[\sum_{i=1}^{a} \frac{\left(\sum_{j=1}^{b_i} \sum_{k=1}^{r} X_{ijk}\right)^2}{rb_i}\right] - CF$
- Within A_i sum of square (SS_{Wi}) $= \frac{\sum_{j=1}^{b_i} \left(\sum_{k=1}^{r} X_{ijk}\right)^2}{r} - \frac{\left(\sum_{j=1}^{b_i} \sum_{k=1}^{r} X_{ijk}\right)^2}{rb_i}$
- Error sum of square (SS_E) $= TSS - SS_T - SS_R$
- Mean square of ith source (MS_i) $= \frac{SS_i}{DF_i}$
- F calculated for i^{th} source (F_i) $F_{[DF_i,\, DF_E]} = \frac{MS_i}{MS_E}$
- Calculate P for i^{th} source with the help of Excel. Type "=FDIST (F_i, DF_i, DF_E)"

Put all the values in the ANOVA table as shown in Table 6.13.

Table 6.13 The ANOVA Format for Two-Factor Nested RBD

Source	*Degrees of freedom (DF)*	*Sum of squares (SS)*	*Mean sum of squares (MSS)*	*F*	*P*
Replication	*r–1*	SS_R	MS_R	F_R	P_R
Treatment	*t–1*	SS_T	MS_T	F_T	P_T
Between A	*a–1*	SS_A	MS_A	F_A	P_A
Within A_i	b_i*–1*	SS_{Wi}	MS_B	F_B	P_B
Error	*(t–1)(r–1)*	SS_E	MS_E		
Total	*tr*–1	TSS			

If $P < 0.05$, then H_0 is rejected and H_1 is accepted for any source such as replication, treatment, A and within A's, and that effect is said to be significant. To identify the difference between any two-treatment mean, apply the post-hoc test. The most common post-hoc test is the LSD test. The formula of SE_d is different for different sources.

SE_d for the difference between any two treatment means or within A_i $SE_{d_B} = \sqrt{\frac{2MS_E}{r}}$

SE_d for the difference between A_i and A_j means $SE_{d_A} = \sqrt{\frac{MS_E}{rb_i} + \frac{MS_E}{rb_j}}$

$CD_i\ 5\% = SEd_i \times t_{DF_E}\ 5\%$ and $CD\ 1\% = SEd_i \times t_{DF_E}\ 1\%$

CD 5% is most common. If more precision is required, it can be at 1% or less. The two-treatment means having a difference greater than or equal to CD differs significantly. On the basis of significance, alphabets may be assigned to different treatment means for easy understanding. Two-treatment means having non-significant differences assigned the same alphabets or different. **Some important points of nested design:** The two-factor nested CRD/RBD design can be used for any experiment when the experimenter is interested in knowing the effect of factor A and, effect of B within A. Some of the important points about Nested RBD are:

- Members of factor A may have varying numbers of b, that is, different drugs may have different concentrations, and the number of concentrations may also be different.

- Different concentrations of factor A are considered as different treatments.
- Treatment SS is split in A and within A's SS.
- The ab_i treatment combinations are evaluated in any basic experimental design such as CRD or RBD depending upon the requirement of experimental conditions.
- Apart from the hypothesis of basic designs, treatment in CRD and replication and treatment in RBD other hypotheses tested in ANOVA are:
  - H_0: No significant difference between A and within A_i.
  - H_1: Significant difference between A and within A_i.
- Difference between two means is tested with respective CD.
- If two means having a difference more than or equal to CD the difference is significant.

There are a number of advanced designs available for different experimental conditions but those are beyond the scope of this book.

Some Important Points

- Design of experiment is a research method to assess the outcome variable related to the treatments on the experimental units under controlled conditions in such a way that the effect of each source of variation in the outcome variable can be estimated separately.
- The basic principles of this method include randomization, replication and local control.
- The intervention or object of assessment is termed treatment.
- **Experimental Material/Unit**: Material unit on which experiments are conducted. Patients/plants/animals/crop fields, etc.
- **Experimental Error:** This variation in outcome variable values due to uncontrolled factors is called experimental error.
- **Randomization:** Random allocation of treatments on experimental units for the validity of statistical tests or to avoid any biases in the experiment.
- **Replication:** Repetition of the same treatment to estimate experimental error based on which final decisions are taken.
- **Local Control**: Arrangement of experimental units into homogenous groups/blocks so that experimental units within the block are homogenous.
- **Precision of Design**: It is the ability with which a design detects the small real difference between treatments.
- **Degree of Freedom (DF):** It is the difference between the number of observations used for the analysis and the number of independent constraints. For a design with 'k' treatment, 'r' replication and 'kr' total observations '$kr-1$' is the DF for total; '$k-1$' is the DF for treatment '$r-1$' is the DF for replication and $[(kr-1)-(k-1)-(r-1)]$ is the DF for error. The sum of DF of different sources is equal to the total DF.
- **ANOVA- Analysis of Variance:** The total variance in the set of observations is split into different components such as treatment, replication, error, etc.
- **F test:** The F value is the ratio of variances. Generally, treatment variance is compared with error variance.
- **Critical Difference (CD)**: If the null hypothesis is rejected by using ANOVA (F test), it implies that all treatment effects are not the same. Then naturally the researcher would like to know which combinations of treatment differ significantly which can be known by post-hoc test. One of them is the CD. It is done by using standard error of the difference of treatment means using the formula SE of difference between means (SEd) $\sqrt{\frac{2V_E}{r}}$. The critical

difference is the least significant difference above which the treatment mean differences are statistically significant. The CD is calculated as, $CD = SE_d \times t_{[DF_e,\alpha]}$.

- The basic designs of field experiments include completely randomized block design (CRD), randomized block design (RBD) and Latin square design (LSD).
- In CRD, the entire experimental area/units are to be homogenous. CRD experiments are possible even for an unequal number of replications. All treatments including control are randomly assigned to the experimental units. For the purpose of data analysis, the observations are arranged according to treatments and replications.
- The CRD design can be used for nutritional experiments like the effect of different products on weight gain by children if an adequate number of children with the same age and initial body weight are available.
- In CRD, all experimental units ('n' units) are homogenous; there are fixed number of treatments t for comparison; when number of replication r for each treatment is same, then $r = n/t$ (number of replication can be different also); each treatment is applied randomly to the experimental units; analysis under missing observation is possible; provides maximum DF for error; H_0: There is no significant difference between treatment effect and H_1: Treatments effects are significantly different
- In RBD, experimental units are divided into blocks with homogenous units within each block. Each block is used as a replication and all the treatments will appear in a block. The total variability is divided into treatment, block/replication and error. The number of blocks is equal to the number of replications and the number of experimental units in each block is equal to the number of treatments. Treatments are randomly allocated to experimental units in each replication.
- In RBD, the total variability is split into three sources, replication, treatments and error; the null hypothesis in RBD is H_0: No significant difference in the effect of treatments and no significant difference between replications.
- LSD is a basic design where entire experimental units are divided into horizontal (rows) and vertical (columns) blocks having homogenous units in both ways. In nutritional trials, the children can be horizontally grouped according to initial body weight and vertically grouped according to age.
- In LSD, the total variability is split into four *viz.,* rows, columns, treatments and error; the null hypothesis H_0: No significant difference in the effect of treatments between rows and between columns.
- Factorial design is used to assess variations in two or more factors simultaneously through the same experiment. When a combination of two drugs (A and B), each with varying doses a_1 and a_2 for A and b_1 and b_2 for B having combination a_1b_1, a_1b_2, a_2b_1 and a_2b_2 will be considered as separate treatments and the experiments are to be planned accordingly instead of conducting two separate experiments for A and B.
- When the levels of factor B vary with levels of factor A, in a factorial design it is called nested design. For example, in health science dosages vary with drugs and in agriculture, levels vary with fertilizers are evaluated in nested designs. In this design, sources are between A and between B within A_1, A_2, A_3, etc. The interaction effect is not there.

Suggested Readings

Huynh, H., and L. S. Feldt, 'Estimation of the box correction for degrees of freedom from sample data in the randomized block and split plot designs', *Journal of Educational Statistics* 1: 69–82, 1976.

Kish, L., *Statistical design for research,* John Wiley & Sons, New York, 1987.

Scheffe, H., *The analysis of variance*, John Wiley & Sons, New York, 1959.

Winer, B. J., *Statistical principles in experimental design*, 2nd ed., McGraw-Hill, New York, 1971.

7 Non-Interventional Observation Research Designs

7.1 Non-Interventional Observation Research

Non-experimental/non-interventional observation research also known as observational research is a major research method in which researchers make use of non-interventional observations related to subjects of study for descriptive, explorative, correlative, analytical types of research. Non-interventional observation method is applicable in many areas, both for qualitative and quantitative research. It is applicable in social sciences, humanities, marketing, business, agricultural sciences, health sciences and many other areas. The broad methods of non-interventional observation research include structured and naturalistic observations on subjects of study by the researcher or through the collection of observational data of selected study participants. In the context of a quantitative framework, observation means data in numerical numbers collected through data collection formats, questionnaires, interviews, etc. Researchers make no external or internal manipulations on the subjects of study as part of non-interventional observation research. In fact, there are situations where study variables cannot be manipulated due to one or the other reasons and observational research is the right option in such cases.

The natural occurrence of events without any external factors or any deliberate control of events in any form is the basis of non-interventional observational research. Observational data collected through a suitably structured data format based on questionnaires, interviews, medical test reports, OPD/IPD records, etc. are analyzed for inferences and to arrive at a conclusion. There are specific designs befitting different situations under the non-interventional observation method of research. Non-interventional studies, while providing the status of a situation or inter-variable relationships under study, also pave the way for further research by developing meaningful hypotheses.

7.2 Non-Interventional Observational Research Designs

Under the non-interventional observation research method, observations are collected from the selected units of study. The units of study may relate to a typical case, a series of cases or a randomly selected sample from the population at a point in time or at different time intervals. The observations from a typical case or series of cases in any subject matter area can form the basis for evolving new concepts or principles in a subject matter area. Similarly, a detailed description of a particular situation or event is often required for follow-up actions or policy interventions in areas like social sciences, humanities, business, commerce, agriculture, health sciences, etc. In some situations, the cause-effect relationship may be of prime importance to take corrective steps or to promote the state of affairs by objectively assessing the existing situation. The applied observational research forms the basis for such actions.

DOI: 10.4324/9781003527183-7

In health sciences, most of the research in para-clinical and basic science areas like anatomy, physiology, biochemistry, microbiology, pathology, forensics, community medicine, etc. and also some studies in clinical subjects are based on observations related to patients as recorded in OPD, IPD, laboratories, operation theatres and even after hospital discharge. The observed changes in physiological functions, anatomical factors and pathological and biochemical parameters in relation to various morbidities/disabilities are the basis of research in these areas. The observational research can be descriptive or analytical in nature. The research designs under observational research are:

- Descriptive Research Case Reports
- Descriptive Research Case Series
- Descriptive Cross-Sectional Research
- Descriptive Longitudinal Research
- Analytical Cross-Sectional Research

These designs are used to identify and describe the situation, relationships, perception, awareness, behaviour, attitude, etc. of people. It can also be used to describe the incidence and prevalence of various diseases and the frequency of occurrence of a phenomenon or to explain the relationship among observable variables related to the subjects.

7.2.1 Descriptive Research Case Reports

A descriptive case report is an in-depth study based on a single unique case, rare in nature with complete details of the case. It may be a rare business case with specific features, a medical case with a rare pattern, an extreme level of profitability/loss of a firm, etc. The researcher makes possible interpretation of the outcome with possible causes for it. Such cases pave the way for further detailed research in this direction.

7.2.2 Descriptive Research Case Series

If the researcher has a series of rare cases in any area of study, it forms the basis for descriptive case series. It may be a group of patients having similar rare clinical features, a group of business firms showing extraordinary profit/loss, or a group of students giving rare types of performance, etc. The observations made are explained without any comparison with control or other cases. It paves the way for further in-depth research based on the hypothesis developed from the case series.

7.2.3 Descriptive Cross-Sectional Research

The descriptive cross-sectional research is a widely used research design under non-interventional observation research method. A cross-sectional random sample representing the population under reference is selected and the required information is collected from the selected respondents to estimate the population parameters. The estimated statistic is tested for statistical significance. In areas like economics, sociology, agriculture, the health sector, business, commerce and many others, the detailed description in parametric form for the existing situation is often required for identifying the areas for policy and development interventions, corrective steps needed for upgradation of the situation, etc. In the health sector, the occurrence of a disease in relation to associated risk factors at a given point in time for a specified population may be of prime importance. The disease prevalence in epidemiology can be assessed when the present status of a disease is the focus of such studies. In cross-sectional studies the exposure and outcome are measured at the same point of time without any time lag. It is less expensive and less time-consuming compared to other methods. It can cover a large-sized population with a scope for disaggregated outcome analysis according to age, sex, etc.

7.2.4 Descriptive Longitudinal Research Designs

Data from study subjects are collected over an extended time period and analyzed to assess changes over time of a phenomenon. These include follow-up studies, trend studies, etc.

7.2.5 Analytical Cross-Sectional Designs

It is a major design under non-interventional observation research methods. Apart from the magnitude, variability and mutual association among dependent variables, the interrelationship of independent variables can also be studied under this design.

7.3 Other Observational Research

Some of the specific observational cross-sectional studies based on study objectives are as under:

7.3.1 Univariate Cross-Sectional Study

In statistics, univariate is used to describe the type of data which consists of only a single characteristic or variable. It is used to assess the frequency of a specific event/disease in relation to a specific situation. It is also applicable to identify perception, awareness, behaviour, attitude, knowledge and practice related to an event/disease. It is also applicable to assess the prevalence and incidence of diseases.

Example: Wages of daily workers across regions, the occurrence of disease across age classes.

7.3.2 Exploratory Research Design

Exploratory research design is used in both qualitative and quantitative research. It is used to understand more about a particular topic or event of interest. This design is applicable when the extent of cases and factors attributable to it are to be assessed for a defined population at a given time. Univariate data can be better perceived using graphs, diagrams, measures of central tendency, measures of dispersion, etc.

Example: Factors influencing the mental health of students, effect of online classes for primary school children, etc.

7.3.3 Comparative Research Design

It is used to compare a common phenomenon in two populations using separate samples from each population. The phenomenon may include mean values, proportions, or scores for knowledge, perception, practices, attitudes, prevalence, incidence, etc. Researchers make attempts to describe similarities as well as differences between groups. Measures of central tendency and dispersions can be worked out and compared. Test of significance can also be applied to assess statistical significance in difference of parameters.

Example: Difference in wages of male and female workers, prevalence of diseases in rural and urban areas, etc.

7.3.4 Co-Relational Research Design

It is a type of non-experimental observation research method used to study the strength of the relationship between two or more variables in natural settings without manipulation or control. The correlation between variables reflects the magnitude of the strength of association as well as the direction of association as positive or negative between variables.

Example: Study time and marks in examination, salt intake and hypertension, food intake and body mass index, physical work hours and body mass index, etc.

7.4 Epidemiological Research

There are three components of epidemiological studies:

- Diseases frequency (disease prevalence, incidence, etc.).
- Diseases distribution (distribution of diseases with respect to place, person and time).
- Diseases determinants (causes or risk factors of diseases).

7.4.1 Measurements in Epidemiology

Epidemiological studies are mostly community-based studies on the determinants and distribution of diseases among people in a specified area at a given time. Hence some of the demographic statistics and vital statistics apply to epidemiology also. Some of the widely used measurements/ indicators in these areas are:

- Measurement of mortality.
- Measurement of morbidity.
- Measurement of disability.
- Measurement of fertility.

These measurements are either in the form of rates, ratios or proportions.

- Rates are calculated as $\frac{a}{a+b} \times 1000$. For example, death rate, birth rate, etc.
- Ratios are calculated as $\frac{a}{b} \times 1000$. For example, sex ratio, doctor-patient ratio.
- Proportions are calculated as $\frac{a}{a+b} \times 100$. For example, proportionate death of infants/mothers.

In epidemiology, all such rates and ratios are with respect to a place and a time as it goes on changing with the change in time or place or both.

7.4.2 Measurement of Mortality (Per Thousand)

- Crude Death Rate (CDR) $= \frac{\text{No. of total deaths}}{\text{Total population}} \times 1000$
- Age Specific Death Rate $= \frac{\text{No. of death in an age group}}{\text{Population in that age group}} \times 1000$

- Diseases Specific Mortality Rate $= \dfrac{\text{No. of death due to a diseases}}{\text{Total patients of that disease}} \times 1000$
- Sex Specific Mortality rates $= \dfrac{\text{No. of deaths of aparticular sex}}{\text{Total persons of that sex}}$
- Infant Mortality Rates (IMR) $= \dfrac{\text{No. of deaths of infants}}{\text{Total number of live births}} \times 1000$
- Child Mortality Rates (CMR) $= \dfrac{\text{No. of deaths of children}(1-4\text{ yrs})}{\text{Total number of children}(1-4\text{ yrs})} \times 1000$
- Under 5 years of age Proportionate Mortality Rate $= \dfrac{\text{No. of deaths of under 5 years children}}{\text{No. of total deaths}} \times 1000$
- Adult Mortality Rates (AMR) $= \dfrac{\text{No. of deaths of adult}}{\text{Total adult population}} \times 1000$
- Maternal Mortality Rates (MMR) $= \dfrac{\text{No. of deaths of mothers}}{\text{Total number of live birth}} \times 1000$
- Case Fatality Rate (CFR) $= \dfrac{\text{No.of deaths due to a disease}}{\text{Total number of that case}}$

7.4.3 Measurement of Morbidity

It refers to the community's position with respect to diseases.

- Incidence Rate (IR) $= \dfrac{\text{No. of new cases of a disease}}{\text{Total population at risk}} \times 1000$
- Prevalence Rate (PR) $= \dfrac{\text{No. of new and old cases}}{\text{Total population at risk}} \times 1000$

7.4.4 Measurement of Disability

Disability rates are of two types: (i) Event type indicators which include the number of days of restricted activity as bed disability days and work loss days; (ii) person type indicators which include limitation of mobility and limitation of activity.

Disability-adjusted life years (DALY) is the sum of years of life lost (YLL) + years lost due to disability (YLD).

$$\text{DALY} = \text{YLL} + \text{YLD}$$

7.4.5 Measurement of Fertility

Relates to the actual bearing of children.

- Crude Birth Rate (CBR) $= \dfrac{\text{No. of live births}}{\text{Total population}} \times 1000$
- General Fertility Rate (GFR) $= \dfrac{\text{No. of live births}}{\text{No. of women}(15-44\,\textit{years age})} \times 1000$

- Age Specific Fertility Rate (ASFR) $= \dfrac{\text{No. of births by women of specified age group}}{\text{Total number of women of that age group}} \times 1000$

7.5 Epidemiological Studies

These are studies related to distribution and determinants of diseases with respect to people, place and time. Observational and analytical studies are carried out.

7.5.1 Descriptive Observational Studies

Such studies are used to describe the distribution of diseases with respect to time (short, medium and long-term fluctuations), place (international, national, regional and local) and person (description of situation, problem, status).

- Cross-sectional (place, persons or time).
- Example: Spread of dengue at national, state, district, block or village level at a given time.
- Distribution of dengue infection according to rural-urban, age, sex, marital status, occupation, social groups, etc.
- Time series (short, medium, long-term) assessment of the situation.

The outcome is new information or new knowledge.

7.5.2 Analytical Observational Studies

It is used to confirm the determinants or risk factors for the diseases.

(i) **Case-Control Study**

- Cases are available with the researcher and the risk factor is required to be confirmed/identified.
- Case/disease group and comparable control group except for disease formed randomly.
- Information on exposure to suspected risk factors ascertained from each member of both groups.
- Exposure rate is calculated for both the groups and compared.
- If the exposure rate of the case > exposure rate of control, implies the exposure leads for the case.
- Odds ratio is worked out to confirm the effect of exposure to risk factors.
- Used to assess the association between risk factors and diseases.
- For example, 'Lung cancer is due to smoking cigarettes'.
- Two groups of cases, the cancer group and the control group are to be formed randomly.
- The control group is the disease-free group (without lung cancer), but comparable with the disease group except for the disease (age, sex, socio-economic, etc.).
- Study proceeds backwards with a history of smoking cigarettes for each selected member in both groups and the frequency of smokers is ascertained.
- Outcome is effect-cause association.

Analysis of case control method is shown in Table 7.1.

Table 7.1 Analysis of Case Control Method

Exposure to risk factor	*No. of cases/with disease*	*No. of control/without disease*
Yes	a	b
No	c	d
Total	a + c	b + d

If $\dfrac{a}{a+c} > \dfrac{b}{b+d}$ then exposure is associated with the case.

Odds Ratio = ad/bc OR can be = 1, or > 1, or < 1

Odds Ratio (> 1) gives the strength of the association of exposure to risk factors.

(ii) **Prospective Cohort**

Cohort means a homogenous group.

- Prospective cohort/longitudinal study/incidence study, forward-looking study, follow-up study, etc.
- Specific risk factor is present in an area and how far it causes the people of the area to have a particular disease is the topic of investigation.
- The researcher begins with a randomly selected exposure group and comparable control group and continues to monitor for future having or not having a disease in individuals belonging to both groups.
- Incidence rates in the exposure group (IREP) and control group (IRCP) worked out and compared.
- If IREP > IRCP, the association of suspected risk factor with the disease is implied or the relative risk (RR = IREP/IRCP) > 1, implies the association of risk factor for the disease.
- Risk difference (RD) = IREP-IRCP
- Attributable risk (AR) = RD/IREP ={(IREP-IRCP)/IREP} × 100.
- Absolute risk is the ratio of total disease in the combined group of exposure and control to the total number of persons in the exposure and control groups.
- Used to assess the association between risk factors and outcome (disease).
- Example, 'Smoking leads to lung cancer'.
- Two groups having comparable features—One exposed to smoking and the other not exposed to smoking are selected.
- Subjects are followed up with periodical medical examination for lung cancer for a specified period.
- At the end, the number of cases in both groups are noted. The outcome is cause-effect association.

The analysis of a prospective cohort study is shown in Table 7.2.

Table 7.2 Analysis of Prospective Cohort Study

Cohort group	*Outcome of disease*		*Total*
	Yes	*No*	
Smoking group	a	b	a + b
Non-smoking group	c	d	c + d
Incidence in smokers	$\dfrac{a}{a+b}$		
Incidence in non-smokers	$\dfrac{c}{c+d}$		

If the incidence in smokers > incidence in non-smokers, then the risk factor is confirmed for having an effect for having the case.

$$\text{Relative Risk}(\text{RR}) = \frac{\textit{Incidence rate in exposure group}}{\textit{Incidence rate in control group}} = \frac{a}{a+b} \div \frac{c}{c+d}$$

The situations are RR < 1; RR = 1 and RR > 1. If RR is greater than one, it implies a much higher risk of incidence of diseases due to exposure to risk factors.

(iii) **Retrospective Cohort Study**

We start with two groups, one representing exposure and the other representing non-exposure groups.

Based on two sets of samples, the occurrence of cases in the past in both the groups will be ascertained.

The results on number of cases under each category are presented in a 2 × 2 contingency table as shown in Table 7.3.

Table 7.3 Results of Retrospective Cohort

Risk factor	*Cases*	*Control*	*Total*
Present	a	b	a + b
Absent	c	d	c + d
Total	a + c	b + d	N

$$\text{Odds for risk factor present} = \frac{a}{a+b} / \frac{b}{a+b} = \frac{a}{b}$$

$$\text{Odds for risk factor absent} = \frac{c}{c+d} / \frac{d}{c+d} = \frac{c}{d}$$

$$\text{Odds ratio} = \text{a} / \text{b} \div \text{c} / \text{d}$$

$$= \text{ad} / \text{bc}$$

or > 1 implies an increased probability of having diseases for persons with risk factors, < 1 implies a decreased probability of having diseases for persons with risk factors and = 1 implies no association of risk factors with disease.

7.5.3 Experimental Studies

In epidemiology, interventional studies such as randomized controlled trials are carried out. Two groups—study group and control group are formed randomly. Intervention (application of drugs or withdrawal of risk factors) is made in the study group. No intervention is made in the control group. After experimentation, the outcome is compared in both groups. The experiment is under the control of the researcher. It involves cost, time and ethics. There are three types of experimental studies in epidemiology.

Clinical/Randomized Controlled Trial

- Allocation of units is made to "Study Group" and "Control Group" randomly.
- Units of the Study Group receive the intervention which can be a new drug or a new procedure.
- Steps in RCT:

(i) Forming a protocol.
(ii) Specifying reference population.
(iii) Random allocation of units to experimental and control groups.
(iv) Intervention in the experimental group.
(v) Follow up.
(vi) Outcome assessment.
(vii) Reporting.

Steps Defined

(i) Protocol: Study aim, two groups, sample size, intervention.
(ii) Reference population: The population where results are applicable.
(iii) Randomization: Selection of study and control groups.
(iv) Intervention: New identified intervention in study group and statuesque (placebo) of control group.
(v) Follow up: Examination of both the groups as required.
(vi) Elimination of bias (if needed).
(vii) Single blind trial-bias of respondents.
(viii) Double blind trial-bias of respondent and researcher.
(ix) Triple blind trial-bias of the preceding + statistician.
(x) Outcome assessment: The results are tested and compared.

7.5.4 Field Trials

- Healthy people form the units of study:

  (i) Preventive trial: Preventive trial (vaccine) on study group with a trial-free control group. Attack/incidence rate of vaccine-preventable diseases assessed for both the groups.
  (ii) Risk factor trial: This trial is with the elimination of risk factors of the study group while the control group is continued with risk factors. Incidence rate of risk factor-prone disease is assessed for both groups.

7.5.5 Community Trial

- In community trials no separation of sick persons or healthy persons is made.
- The entire community is divided into two groups, one group receiving the intervention and the other not receiving the intervention (may be an educational programme).
- Statistically significant reduction in the incidence of problems in the study group proves the positive impact of the intervention.
- Example: Anganwadi visits versus malnutrition.

7.5.6 Non-Randomized Trials

- Uncontrolled trials: Evidence from rare study groups is compared with historical evidence.
- Natural experiments: Natural situations like earthquakes, famines, floods, etc. are treated as experiments.
- Before and after comparison trials: Before interventional evidence of the group receiving the (ii) control cohort intervention is treated as control.

7.5.7 Screening Tests

- Screening for diseases among people in a community is a step toward prevention of diseases.

- Screening for diseases means searching for the presence of diseases/defects in apparently healthy individuals by means of rapidly applied tests/medical examinations.
- Screening tests are different from diagnostic tests.

The sequence of screening tests for disease is given in Figure 7.1.

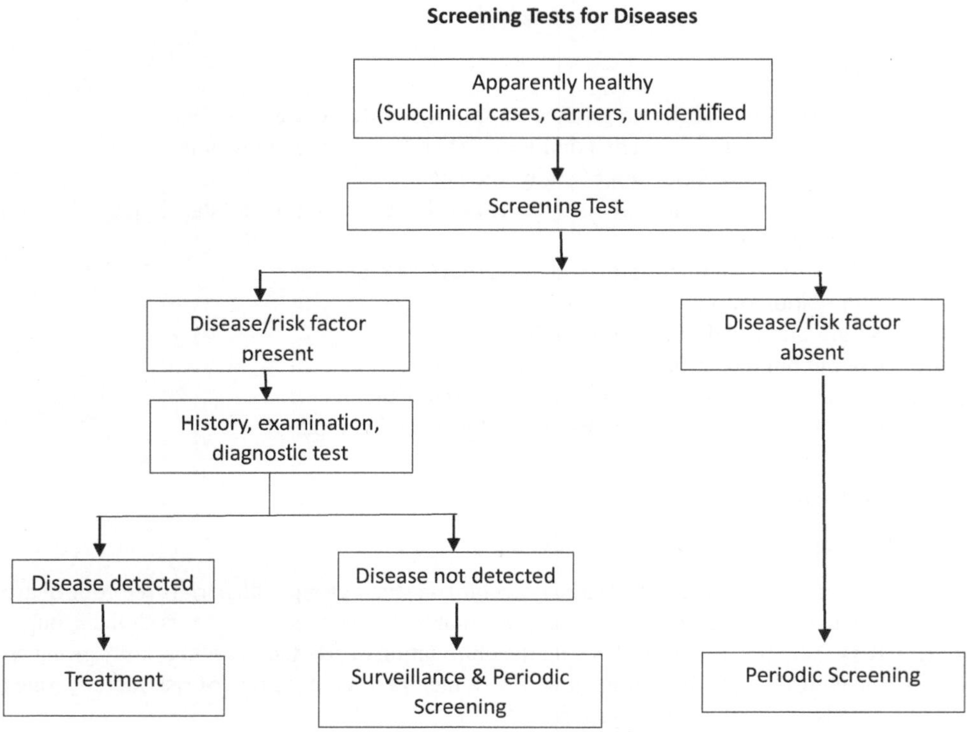

Figure 7.1 Sequence of Screening Test for Disease

Purpose of screening test

- Case detection (neonatal screening for breast cancer).
- Control of disease (screening of immigrants for communicable diseases).
- Research purposes.

Types of Screening

- Mass screening (all school children for malnutrition).
- Selective screening (screening of immigrants for syphilis).
- Multi-purpose screening (two or more tests at one time—tests for syphilis and HIV).

Criteria of Screening Tests

- Acceptability (to people/community).
- Repeatability (scope to repeat, if required).

- Consistency (more or less the same results when repeated).
- Validity (ability to distinguish problematic/diseased from non-problematic/non-diseased).
- Both sensitivity and specificity is expressed in per cent.
- When alternative tests are available for screening for diseases, the tests with higher sensitivity and specificity are used.
- Sensitivity of a screening test is its power to identify the disease correctly.
- Specificity of a test is its power to identify the non-disease correctly.
- For the calculation of sensitivity and specificity, two groups are required.

Group A, those having and Group B, those not having the disease.
A summary of the results of screening test is shown in Table 7.4.

Table 7.4 Summary of Results of Screening Test

Screening Test Results	*Diseased*		*Total*
	Yes	*No*	
Positive	a	b	a + b
Negative	c	d	c + d
Total	a + c	b + d	a + b + c + d

Notes: 'a' True +ve 'b' False +ve
'c' False −ve 'd' true −ve
Sensitivity = (a/a + c) × 100
Specificity = (d/b + d) × 100
Predictive value of +ve test = (a/a + b) × 100
Predictive value of −ve test = (d/c + d) × 100
Percentage of false +ve test = (b/b + d) × 100
Percentage of false −ve test = (c/a + c) × 100

Some Important Points

- Natural occurrence of events without any external factors or any deliberate control of events in any form is the basis of observational research.
- Observational data collected through suitably structured data format based on interviews, medical test reports, OPD/IPD records etc. are analyzed to have findings and arrive at a conclusion.
- Most of the research in para-clinical and basic science areas like anatomy, physiology, biochemistry, microbiology, pathology, forensics, community medicine, etc. and also some studies in clinical subjects are based on observations related to patients as recorded in OPD, IPD, laboratories, operation theatres and even after hospital discharge.
- **Descriptive Research Case Reports:** It is an in-depth descriptive study based on a single unique case, rare in nature with complete details of the diagnostic test conducted, its results, treatments made and its responses, etc.
- **Descriptive Research Case Series:** A group of patients having similar clinical features along with diagnosis and interventions made are observed for a period of time. The observations so made are explained without any comparison with control or other cases.
- **Descriptive Research Cross-Sectional Studies:** It falls under descriptive observational study. It is used when the occurrence of a disease is related to associated factors at a given point in time for a specified population.

- **Exploratory Research Design:** This design is applicable when the extent of cases and factors attributable to it are to be assessed for a defined population at a given time.
- **Comparative Research Design**: Used to compare a common phenomenon in two populations using separate samples from each population. The phenomenon may include knowledge, perception, practices, attitudes, prevalence, incidence, etc.
- **Co-Relational Research Design**: It is used to study the strength of the relationship between two or more variables in natural settings without manipulation or control.
- **Developmental Cross-Sectional Research Design**: Data at one point in time are collected and analyzed to assess the developmental process.
- **Developmental Longitudinal Research Design:** Data collected over an extended time period and analyzed to assess changes over time of a phenomenon. These include follow-up studies, trend studies, etc.
- **Epidemiological Descriptive Observational Studies**: Such studies are used to describe the distribution of diseases with respect to time (short, medium and long-term fluctuations), place (international, national, regional and local) and person (description of situation/problem/status).
- **Analytical Observational Studies**: It is used to confirm the determinants of the diseases.
- **Case-Control Study:** Cases are available with the researcher and the risk factor is required to be identified, case/disease group and comparable control group except for disease formed, information on exposure to suspected risk factor ascertained from each member of both the groups, exposure rate is calculated for both the groups and compared, If exposure rate of case > exposure rate of control implies the exposure is a risk factor for the case, it is the same as a retrospective cohort where odd's ratio is worked out.
- **Prospective Cohort:** The researcher begins with the exposure group and comparable control group and continues to monitor for future having or not having disease in individuals belonging to both groups. Incidence rates in the exposure group (IREP) and control group (IRCP) are worked out and compared.
- Screening for diseases means searching for the presence of diseases/defects in apparently healthy individuals by means of rapidly applied tests/medical examinations.

Suggested Readings

Park, K., *Park's textbook of preventive and social medicines*, Banarsidas Bhanot, Jabalpur, 2013.

Sharma, B. S., *Research methods in social sciences*, Sterling Publishers Private Ltd., New Delhi, 1983.

Suryakantha, A. H., *Community medicine with recent advances*, Jaypee Health Sciences Publisher, New Delhi, 2017.

William, A. O., *Epidemiology concepts and methods*, CBS Publishers and Distributors Pvt. Ltd., New Delhi, 2008.

8 Survey, Interview and Online Research Designs

8.1 Introduction

The survey (offline), interview and online survey methods of research have a wide range of applications in various areas of research. The online surveys that evolved in recent years have added to the scope and applications of survey methods. Under these methods of research, the data is collected from selected study respondents by seeking opinions and responses to the items identified as per the objectives of the study. The responses of respondents can be information in numerical form, perception, insights and opinions as multiple-choice questions, dichotomous questions, open-ended questions, close-ended questions, scale-based marking, etc. These methods are widely used in social sciences, humanities, commerce, marketing research, business studies, economics, sociological studies, education, law, community-based and other health studies and many other areas.

8.2 Survey Research

Survey research includes the use of schedules, questionnaires, forms and other survey-based tools to gather data/information from the respondents about their perceptions, opinions and ideas. After survey data is gathered, statistical analysis is performed to produce insightful study findings and recommendations. Survey research is one of the most effective research methods in many areas of study. Online surveys are becoming a very popular approach these days to get data from individual persons or a group of people. They are prepared as pre-structured survey questions meant to encourage participants to reply with a minimum time lag. Many companies use survey research to gather precise data based on respondents' perceptions. Conventional survey research uses both qualitative and quantitative techniques to gather data, opinions, perceptions and other information from a set of respondents through a series of survey questions. It is usually the initial step in starting more comprehensive and lengthier qualitative or quantitative research procedures, such as focus groups, on-call interviews, surveys, polls as well as the process of quickly obtaining information about popular subjects. To further their research, researchers may use a combination of quantitative and qualitative survey approaches in a number of situations.

8.2.1 Modes of Survey Research

The choice of survey research design depends on two important factors—the tool used in survey research and the amount of time needed to carry out the study. Depending on how survey research is conducted, there are three main survey research designs.

DOI: 10.4324/9781003527183-8

Online/Email: One of the most often used survey research techniques available today is online survey research. Research based on Internet surveys is inexpensive, conducted quickly and yields responses that are relatively reliable. Data gathering and compilation are quicker and easier using Google Forms.

Phone: Data from a wider range of the target respondents can be gathered through computer-aided telephonic interviews (CATI) or survey research performed over the phone. It's likely that phone surveys may cost more money to conduct and take more time than other methods. However, the time and cost for transport of the researcher/representative for face-to-face conduct of the survey is saved.

Face-to-face: When it is possible to meet with respondents one-on-one, researchers conduct in-person, comprehensive interviews. This approach has the highest response rate, although it can be expensive and time-consuming. Surveys on subjects where data privacy is important, and which seek more accurate data are better suited for this method of data collection.

Based on the time taken, survey research can further be classified into two methods:

Longitudinal Survey Research

In longitudinal survey research, surveys are administered over a broad spectrum of time periods. Both qualitative and quantitative data may be gathered over time utilizing this survey research method. The behaviour, preferences and attitudes of the respondents are continually noted and tracked over time to determine the causes of any changes in the respondents' choices or behaviour. Let's say a researcher wants to find out the changes in the eating habits of teens. To make sure the data gathered is accurate, the researcher will need to monitor a sample of teenagers for a significant amount of time.

Cross-Sectional Survey Research

Cross-sectional surveys are used by researchers to acquire information and opinions from a target population at a certain point in time. It is used by organizations like industry, marketing, education, sociology, economics, healthcare, small- and medium-sized companies (SMEs), etc. Analytical or descriptive research can be conducted using cross-sectional surveys. It is rapid and facilitates the accurate collection of data by researchers. When a topic requires analytical or descriptive investigations, researchers turn to the cross-sectional survey research method.

Additionally, survey research is divided into two categories based on the sampling techniques used to choose study samples: non-probability sampling and probability sampling. Under probability sampling, each member of a population should be treated equally when it comes to being included in the survey study sample. Using probability theory, the researcher selects the respondents for this sample technique. Numerous probability research techniques exist, including stratified random sampling, cluster sampling, systematic sampling and simple random sampling. In non-probability sampling, the researcher selects the sample for data collection based on his or her expertise and experience. Convenience sampling, snowball sampling, sequential sampling, judgmental sampling, quota sampling and other non-probability sampling methods are among them.

8.2.2 Sequence of Survey Research

Design Survey Questions: Grammatically and logically sound survey questions can be created using a standard questionnaire or by creating them during a brainstorming session followed by post-validation. It is very important to comprehend the purpose of the survey as well as the anticipated results. In many surveys, learning about respondents' preferences among the alternatives offered is more valuable than knowing the details of their open choices. In these circumstances, a researcher may use multiple-choice or closed-ended questions; on the other hand, open-ended questions may be included in the questionnaire if perception or insights on certain topics are needed. The surveys ought to have a thoughtful mix of closed-ended and open-ended questions. This can be made possible by the use of the Likert scale, Semantic scale, Net promoter score question, etc. to avoid fence-sitting.

Fixing Target Population: The researcher will have to decide the target population, the list of which must be available if a random sample representing the population is intended. A pilot study is advisable if the right sequencing and way of presenting the questions are to be streamlined. Thus, the result generated can be in accordance with the requirement and generalized for the entire population.

Conduct of Survey: Once the survey questionnaire is distributed to the selected respondents, the researcher needs to wait patiently for the feedback. The nature of the target population and the region in which the survey is to be conducted have to be kept in mind while scheduling the survey. Besides the distribution of questionnaires, surveys can also be conducted using social media, emails or website surveys in order to maximize the response.

Survey Data Analysis: Real-time feedback analysis should be done to spot trends using accepted statistical methods. Survey feedback analysis techniques such as conjoint analysis, cross-tabulation, GAPs (difference between actual and expected), TURF (total unduplicated reach and frequency to assess market potential) and many other techniques can be used to identify and provide insight into respondents' perceptions and behaviours. The findings may be used by researchers to assess the level of agreement with a statement, the extent of satisfaction with some services, to evolve remedial measures to raise employee and customer satisfaction, among other things. Research results from economics and sociology are generally used for policy formulations.

8.2.3 Validation of Questionnaire

A questionnaire needs to be validated to ensure its suitability for any specific sample survey.

1. **Validity by experts:** For this, the contents and sequences of items in the questionnaire will have to be gone through by experts who are well-versed with the topic of research so that the effectiveness and completeness of the questionnaire are ensured. Subsequently, the questionnaire is examined for errors/difficulties/confusion in getting true responses from respondents to various questions.
2. **Pilot-testing:** It is done on a subset of the intended population. The size of the sample size must increase with the increase in questions in the questionnaire. It will help to remove irrelevant/confusing questions from the questionnaire. Additionally, the correctness of answers to negatively phrased questions in relation to positively phrased questions can be assessed while entering the data collected under the pilot study. In some cases, reversal of scale values for negatively phrased questions may be required. Necessary cleaning of data can be done.
3. **Principal component analysis (PCA):** It will help to identify underlying components. The component or factor loading will tell as what factors are being measured through the questions. Questions that measure the same thing should load onto the same factors. Factor loadings range from -1 to $+1$. Factor values that are $\pm$ 0.60 or higher can be grouped. In some cases, a question will not load onto any factors. In fact, one must determine what factors represent by looking for common themes in the questions that load onto the same factors. If 'k' factor themes are

identified, it means the survey is measuring at least 'k' things. Finally, questions loading onto the same factors can be aggregated or combined and compared during the final analysis.

4. **Internal consistency of questions loading:** It checks the correlation between questions loading onto the same factor. It is a measure of reliability as it checks whether the responses are consistent. By including some reverse questions having opposite answers of all previous questions, the consistency and vigilance of the respondent can be ascertained. Another standard test for internal consistency is Cronbach's Alpha (CA). The value of CA ranges from zero to one. The value of CA above 0.6 is acceptable and the value above 0.70 is quite alright. The CA value can be improved by dropping one or two questions if the CA value is low.
5. **Revision of questionnaire:** The questionnaire will have to be revised based on information/feedback from PCA and CA. Here also one can retain a question even when it does not adequately load onto a factor, if it is important in the overall framework and it can be analyzed separately.

8.2.4 Factors of a Survey Research

The primary and most important benefit of employing surveys in research is that the researcher may get data on the research hypotheses that have been established for the study. Depending on the intended audience and purpose of the survey, the researcher may pose these questions in a variety of ways. For a study to be appropriately constructed, planned and carried out to produce certain results, the researcher must be clear about the goal of the investigation before beginning to design it. A few things should be considered while planning a survey:

- Main objective of conducting the survey.
- Plan to utilize the data collected.
- Decisions to be taken based on the results of the survey.
- Understand the respondent's views, perceptions and opinions about a topic/product. If the survey is carefully planned, it is possible to get the respondents' views/insights about the problem/topic as well as their suggestions for improvement. The researcher must be really clear about how to secure their responses and how to use the answers in order to encourage them to answer accurately. As the privacy of online and mobile surveys has been established, an increasing number of respondents are comfortable providing their opinions via these platforms.
- Present a medium/source for discussion: A survey acts as a perfect platform for respondents to provide criticism, suggestions and plus points about the topic of study. In areas like marketing, the customers' views on important topics like product quality, customer satisfaction level etc. can be realistically assessed. This can be achieved by including open-ended questions for the respondents to express their thoughts.
- Paves the way for strategic changes: Researchers can use the criticism and feedback received from the surveys to improve the situation/product/services. Once the agency successfully makes the suggested changes/improvements, it can conduct a follow-up survey to measure the impact of changes made as the feedback keeping the initial benchmark results. By doing this activity, the organization can track what was effectively improved and what improvements are still needed.

8.2.5 Survey Research Scales

The four main basic scales for the measurement of variables, nominal, ordinal, interval and ratio scales are also used in survey research. Depending upon the scales used, the nature and type of data analysis in survey research can be planned by the researcher.

8.2.6 Survey Research Stages

Survey research design provides easy access to details at a limited cost. Researchers frequently employ this technique to comprehend and evaluate the current state of affairs and anticipated future demands in relation to a good or service. Information gathering via a carefully planned survey inquiry may be far more efficient and fruitful. There are five stages of survey research design:

- Deciding on the aim of the research.
- Deciding the target population and sample.
- Finalize survey method.
- Designing the questionnaire.
- Collection of data.
- Analyze data for results.

8.2.7 Tips for Survey Research

The key to gathering the data that a researcher needs to make important decisions based on study findings may lie in selecting the appropriate survey design. Selecting an appropriate topic, right questions and suitable design are crucial. Every stage of the process, including developing survey questions, sending out questionnaires by email or links and evaluating the results, is made simple and effective with the use of web-based software like Google Forms.

- **Set the SMART goals:** What is intended to achieve with the survey, how can it be measured promptly and what are the tentative/expected results?
- **Choose the right questions:** Prepare the questionnaire by choosing those specific questions relevant to the research.
- **Begin the questionnaire with generalized questions:** Preferably, start the questionnaire with general information related to the respondent followed by basic questions on the topic of study.
- **Include topic-related questions:** Choose the best, most relevant, limited number of questions (15–20 questions) covering multiple-choice, rating scale, open-ended, yes or no type questions, etc.
- **Pre-testing:** Once the questionnaire is created for the survey, it's time to test it for corrections and changes to make it more user-friendly.
- **Distribution of questionnaire to respondents:** Once the survey questionnaire is ready, it is time to distribute it to the right audience through the selected online/offline media for their compliance.
- **Collect and analyze responses:** Collect the filled questionnaire and enter the data on Excel or any specific worksheet sheet prepared for data entry, with all the necessary categories mentioned.
- **Prepare the report:** Analyze the data and prepare the report in the report format.

8.3 Interview Methods

The important terms used in the interview method of research are:

Investigator/Respondent: The researcher or his/her representative is called the investigator. One who supplies the information or from whom the information is collected is called the respondent.

Interview Schedule: The material/format used to collect the required information for any study is called a schedule. A schedule is generally filled in by the investigators asking questions or through discussion with the respondents.

8.3.1 Designs of Interview Method

The interview-based research can be carried out under different modes or designs. These include:

- Questionnaire-based direct interview method.
- Schedule-based indirect interview method.
- Internet-based Google Form method.
- Telephonic interview method.

    - **Questionnaire-Based Direct Interview Method:** Such studies are made based on a structured questionnaire which includes a set of questions or quarries developed based on the objectives of the study and corresponding data requirements. Normally the responses to the questions are provided by the respondents/subjects selected for the study themselves on request by the researcher or his/her representatives.
    - **Schedule-Based Indirect Interview Method**: A schedule is like a questionnaire, but it is filled by the researcher or his/her representative by interacting personally with the respondent.
    - **Google Form-Based Indirect Interview Method**: With the advent of ICT and access to several social platforms by a large majority of people these days, ICT-based Google Forms are widely used for the collection of research information in a variety of studies. During the COVID era when social distancing and other restrictions were in vogue for one-to-one meetings, Google Forms became a more research-friendly technique for data collection.

To gather information about a topic, open-ended questions are typically used in an interview which is a qualitative research approach. Most of the time, the interviewer is a subject matter expert who uses a carefully thought-out and executed set of questions and answers to understand the opinions of the respondents. When it comes to gathering information from the target market, interviews and focus groups are comparable, but they function quite differently. For example, focus groups are limited to a small group of six to ten people, whereas surveys are quantitative in nature. Interviews are conducted by drawing a sample from the population and are done in a conversational tone.

8.3.2 Types of Interviews

For a study where information can only be gathered by meeting and establishing a personal connection with the target audience, a researcher must go for the interview method of data collection. Using interviews, researchers may stimulate participants and get detailed input that they need. In research, there are three basic kinds of interviews:

Structured Interviews:

Structured interviews are characterized by their incredibly rigid operations and limited scope allowing participants to gather and evaluate data. For this reason, it is sometimes referred to as a standardized interview and works better in quantitative research. The questions of the interview are pre-planned based on the specific details that are needed. Survey research also makes extensive use of structured interviews to ensure consistency across all interview sessions. They can be either open-ended or closed-ended, depending on the kind of intended audience. Open-ended questions may be used to learn more about the respondent's perspective and insight, whereas closed-ended questions can be used to understand user preferences from a list of possible answers.

Semi-Structured Interviews:

Semi-structured interviews preserve the fundamental interview framework while allowing the researcher a great amount of flexibility in questioning the respondents. Researchers are afforded a notable degree of flexibility, even in the case of guided conversations between interviewees and researchers.

As long as the researcher keeps the format in mind and ensures that this kind of research interview does not require numerous rounds, they are free to pursue any idea or creatively utilize the whole interview. Gathering data for a research project always requires further questioning of respondents. When a researcher needs in-depth knowledge on a subject but lacks the time to study, semi-structured interviews work best.

Unstructured Interviews:

Unstructured interviews, also known as open interviews, are generally defined as discussions conducted with the intention of obtaining information for the research project. These interviews are more like a conversation around a central theme and, hence have the least number of questions. Most researchers who use unstructured interviews hope to establish a rapport with their subjects using this method, which increases the likelihood that they will be completely honest in their responses. The researchers can approach the participants in whatever ethical way to obtain the needed information for their research topic because there are no rules to follow. There are no rules for such interviews, therefore, the researcher must be careful to ensure that the respondents stay focused on the primary goal of the study.

The participant's interests and abilities should be the main considerations throughout the interview. The researcher should make all efforts to adhere to the acceptable boundaries of the study, and all discussions should be done inside them. The researcher's expertise and experience should align with the interview's objectives. The dos and don'ts of unstructured interviews should be understood by the researchers.

8.3.3 Designs of Interviews

Depending on the requirement of the research study, any of the following three methods can be used to conduct a research interview.

(i) **Personal Interviews:**
One of the most popular interview formats is the personal interview, in which the interviewer asks the questions directly and in person to the respondents. A researcher might plan his questions such that he can record any remarks or viewpoints that the respondent makes that may be useful for the study. Although doing in-person interviews takes more time, the response rate is generally higher Having the interviewer there to address any

questions, the interviewee may find it more comfortable to give correct responses. However, if the interviewee feels uncomfortable with the interviewer, it could be difficult to receive an honest response.

(ii) **Telephonic Interviews:**
To conduct research efficiently, telephonic interviews are frequently employed and may be easily combined with Internet surveys. Telephonic interviews save money and time and are relatively easy to conduct. However, respondents not picking up the phone from unknown numbers and not answering under the pretext of being busy are some challenges that a researcher may face in the case of telephonic interviews.

(iii) **Email or Web Page Interviews:**
As respondents nowadays move toward a more virtual environment, there is an increasing amount of research being done online, and it is important for researchers to adjust to this shift. The prevalence of Internet connectivity has led to a rise in the popularity of web-based and email-based interviews which are now the most common form of interviews. There is no better tool for this than an online survey. Online shoppers are becoming more and more popular, which makes them a fantastic segment to be able to do interviews that will produce data for informed decision-making.

Choosing the type of interview best suited for data collection for research work depends on the objective of the research.

8.4 Online Survey Research

It is a time- and money-saving method for gathering information for research, which uses Internet connectivity for communication between the researcher and respondents. The old pen-and-paper research methods have been superseded by online research since the invention of the internet. Considering their affordability and convenience of use, online surveys have a far greater influence than traditional methods. Because respondents to internet research are ensured the privacy of their identity, response rates are significantly greater than for other research methods.

The advancement of online survey research is progressing with the development of the internet and social media. In terms of easy and cost-effective database access, online platforms like social media have acted as a catalyst for online research.

A variety of online research tools, including focus groups, questionnaires, polls and forms, are crucial for obtaining online data for research. Small and big organizations may now perform market research with little to no expenditure. Product testing, audience targeting, database mining, customer happiness and many other sectors may all benefit from online research. Because of the accuracy of the results they provide, researchers frequently employ five of these online research methodologies.

(i) **Online Focus Group:** It is a subset of internet research methods that are employed in consumer, political and business-to-business (B2B) research. The task of leading and supervising the focus group falls to the moderator, who extends invitations to pre-selected and qualified participants who fit a certain interest area to participate at a specified time. Typically, participants get incentives for participating in the conversation.

(ii) **Online Interview:** While the needed standard practices, communication with respondents and sample methods are different in this online research approach from in-person interviews, they are still quite similar. A variety of computer-mediated communications (CMC), primarily SMS or email, are used to arrange online interviews. These interviews are divided into synchronous and asynchronous approaches based on the response time.

Asynchronous online interviews take place by email, and the replies are typically not received in real-time, whereas synchronous online interviews are conducted through platforms like online chat. Similar to in-person interviews, online interviews delve into the opinions and thoughts of participants about certain subjects to gain an understanding of their backgrounds, perspectives and dispositions.

(iii) **Online Qualitative Research:** There are other forms of online research, particularly qualitative research, outside the popular online focus groups and online interviews. These forms consist of communities, blogs and mobile diaries. These techniques are very practical for researchers to collect data for their research study and help save money and time. Because respondents may be added through surveys, panels, or already-existing databases, online qualitative research methods offer a higher degree of sophistication than any other traditional approach.

(iv) **Online Text Analysis:** This analysis method is an expansion of text analysis and consists of a compilation of different online research samples used to extract knowledge from information that is accessible online. This method of doing research online allows researchers to provide written, spoken or visual explanations of communication format categories: Documents, sentences, paragraphs, quasi-sentences, web pages, etc. Although it is most frequently employed for quantitative research, researchers also employ qualitative approaches to improve text interpretation.

(v) **Social Network Analysis:** The growing popularity of social networking platforms has led to the acceptance of social network analysis, a new online research method. Graph theory may be used by a researcher to map and quantify flows and interactions between individuals, groups, organizations, URLs and computers through social network analysis.

Some Common Types of Online Research:

(i) **Customer Satisfaction Research:** In the past, this kind of research was done over the phone, but in the modern day, clients are used to receiving emails soliciting their opinions about their latest experiences with businesses. For example, you would want to know about client satisfaction if you are the owner of a newly launched restaurant. After getting their email address, you can either send them the survey or have it filled out after their meal.

(ii) **New Product Research:** The launch of a new product requires an understanding of its chances of being popular with the target audience. To conduct a new product response test, a chosen number of individuals may try the product, and feedback can be gathered instantly.

(iii) **Brand Loyalty Understanding:** Numerous small and large businesses rely only on established brand loyalty to remain in business. It is undoubtedly important, but to preserve or enhance it, every company must work on it. To find out what draws a client to a specific brand or what deters them from being loyal to yours, you can conduct online research.

(iv) **Employee Engagement and Employee Satisfaction:** Success comes from knowing what your workers think about working for your company. To ensure that workers effectively contribute to the company's growth, it is important to regularly monitor their attitude and morale. Surveys ought to be sent to sustain employee satisfaction as well as to increase staff engagement.

Things to Keep in Mind for Online Survey Research:

The following are some points to take care of while designing an online survey for research:

(i) **Avoid Open-Ended Questions:** Open-ended questions require careful consideration before being answered, which might lengthen the completion time. They may become so irritated that they give up on the survey altogether. Respondents will find it much simpler to complete yes-no, multiple choice and rating questions.

(ii) **Show Urgency with Patience:** It is quite OK to send out multiple invites for responders to complete the form if you need a response on a crucial matter. However, your database must be fully aware of this and have no objections to it for this to happen. Above all, after doing an online survey, wait patiently for the results. Assign a team member to oversee the completion of this survey from start to finish.

(iii) **Precise Surveys Produce Better Results:** Surveys can take long sittings for respondents which discourages them from attempting any future surveys you would send out. By using multiple choice questions (MCQs) with precise answers, you may shorten the survey's length, and the time respondents must spend completing it.

Online Research Advantages:

Access to Data Across the Globe: Researchers may use the internet as a sophisticated platform to focus their attention on finding important information that would otherwise take a lot of time to find. Even when they have a deadline approaching, it is rather simple for them to conduct research online.

Minimum Investment of Time and Resources: Online platforms are great places for people to search for information to increase their knowledge. Every day, new information is released, and researchers make use of this to their advantage. It has made publishing and information gathering easier, which has resulted in significant time and cost savings.

Central Pool of Facts and Figures: The best advantage the Internet has to offer is that students can use it when searching for academic content while researchers and statisticians are constantly looking for updated information on a variety of significant issues.

Capable Tool for Collecting Information: To collect or disseminate important data, surveys, quizzes and polls are carried out online using tools like emails, QR codes, embedded content on websites, etc.

Some Important Points

- A survey is the combination of questions, processes and methodologies that analyze data about participants.
- The survey research tool for data collection consists of methods like online/emails, phones, face-to-face interaction. A survey can be longitudinal or cross-sectional.
- Steps in survey research methods are to decide on survey questions, finalize a target audience, conduct surveys via decided mediums and analyze survey results.
- In survey research nominal, ordinal, interval and ratio scale-based data can be collected.
- Privacy/secrecy of respondents and responses possible, applicable for both quantitative and qualitative research.
- The stages of survey methods are: Fix aims and objectives, selecting a sample from a target population, identifying the survey method, designing the questionnaire, conducting the survey, collecting data and analyzing responses.

- Research survey can be made successful by selecting smart goals, choosing the right questions, starting with few simple and general questions, making questions with varying options to answer especially more questions with 'yes or no option', keeping the survey instruments and electronic devices intact, timely distribution of questionnaire to respondents, collection and analysis of data and finalizing the report.
- In interview research, the investigator will have to move directly or indirectly from one subject of study to the other to get the required data through direct or indirect contacts.
- **Interview Schedule/Questionnaire**: An interview schedule is a structure used to gather the data needed for any research. While the respondents often fill out the questionnaire, the investigators typically fill up a schedule by talking with respondents or by asking questions.
- Validity of the questionnaire can be made using (i) face validity for content and sequence by experts, (ii) pilot study on a small group for assessing lack of clarity, if any, (iii) factor loading through principal component analysis to make grouping of factors, (iv) reliability test through Cronbach's Alpha and (v) revision based on feedback from (i) to (iv).
- Interview-based research can be carried out under different modes or designs like the questionnaire-based direct interview method, schedule-based indirect interview method, internet-based Google Form method, telephonic interview method, etc.
- The types of interview research can broadly be grouped as structured interviews, semi-structured interviews and unstructured interviews.
- Structured interviews are also known as standardized interviews and are more useful for quantitative research. Questions in this interview are pre-decided according to the required detail of information.
- While retaining the fundamental interview structure, semi-structured interviews provide the researcher with a great deal of flexibility in questioning the participants.
- Open or unstructured interviews are conversations that are carried out with the intention of gathering information for a research project.
- Research interviews can be conducted in three different ways: In-person, over the phone, or email or a website.
- There are five online research methods- online focus group, online interview, online questionnaire, online text analysis and social network analysis.

Suggested Readings

Bloch, A., C. Phellas, and C. Seale, *Structured methods: interviews, questionnaires and observation in researching society and culture*, 3rd ed., Sage Publications Ltd, London, 2011.

Fowler, F. J., *Survey research methods,* 4th ed., Sage Publications, 2009, https://doi.org/10.4135/9781452230184

Glasow, P. A., *Fundamentals of survey research methodology*, MITRE, Washington, DC; C3 Center, McLean, VA, 2005.

https://researchmethod.net/survey-research/

Jaber, F. G., and J. A. Holstein, *Handbook of interview research*, Sage Publications, 2001, https://doi.org/10.4135/9781412973588

More, J. M., 'Determining sample size', *Qualitative Health Research* 10(1): 3–5, 2000.

Patton, M. Q., *Qualitative evaluation and research methods*, 2nd ed., Sage Publications, Newbury Park, CA, 1990.

Sampson, P., 'Qualitative research and motivation research', in *Consumer market research handbook*, 3rd ed., edited by R. M. Worcester and J. Downham (eds.), Elsevier, Amsterdam, 1986.

Walker, R. (ed.), *Applied qualitative research*, Gower Publishing Company Ltd., Hants, 1985.

Williams, D. G., and N. A. Johnson, *Essentials in qualitative research: a notebook for the field*, McMaster University, Hamilton, 1996.

9 Qualitative Research Designs

9.1 Introduction

In qualitative research, non-numerical information or data, such as verbal or written materials, are gathered and analyzed to examine people's ideas, perceptions, experiences, insights and other aspects of any given research topic. The primary method of collecting data for qualitative research is conversational or open-ended dialogue. This approach looks at 'why' and 'what' individuals think about the things they do. The fields of social science, business studies, marketing research and other fields can benefit more from qualitative research. It focuses on how people perceive, understand and behave in relation to one another. The intent behind the creation of qualitative research methodologies is to throw light on how the concerned audience behaves and perceives a certain issue. Qualitative research methods are also used by researchers to capture in-depth and non-numerical information. Qualitative approaches yield more descriptive results, and the data readily allows for the drawing of conclusions. Social and behavioural sciences are the fields that use qualitative research methodologies. Because of the complexity of today's environment, it might be challenging to comprehend how others think and see situations. As online qualitative research methods are more detailed and expressive, they facilitate understanding of such facts.

9.2 Data Collection Methods

The following are the ways to collect data for qualitative research. The data collection methods in qualitative research broadly include written expression by the participants, observation by the researcher and interactive interviews with the participants by the researcher. The following are some of the widely used qualitative research methods.

9.2.1 Personal Interview

An important approach for doing qualitative research is conducting in-person interviews. The interviews are taken one respondent at a time. This approach is solely verbal and allows for opportunities to get detailed information from the respondent.

The advantage of this approach is that it offers an excellent opportunity to collect accurate information on people's beliefs and motives. An experienced researcher can get insightful data by posing the appropriate questions to the respondents. If the researchers require further data, they should offer follow-up questions that will facilitate the collection of additional information. The interviews can be held in person or telephonically. However, in-person interviews are more effective as the researcher gets to observe the body language of the respondent and can match the responses.

DOI: 10.4324/9781003527183-9

9.2.2 Focus Groups

Focus groups are a frequently used technique for collecting information from a group for qualitative research. A target group of six to ten participants often makes up a focus group. Finding the answers to the question as 'why', 'what' and 'how' is the primary goal of the focus group. The main benefit of focus groups is that the researchers do not always have to speak with the participants face-to-face. Focus groups may now receive online surveys on a variety of devices, and the replies can be collected for additional processing.

Focus group approach is more costly method among different online qualitative research methodologies. They are generally used to clarify intricate procedures. Testing of new products or new concepts and market research can gain benefit greatly from focus groups.

9.2.3 Records and Documents

This approach uses pre-existing, credible documents and related information sources as the data source. Fresh research can get advantage from this data. This is like visiting a library where one can go through books and other publications to get required data that is probably needed in the study.

9.3 Process of Observations

It is a research process that uses subjective methodology to gather systematic information or data. The main purpose of this method is to compare disparities in quality. The five main senses—sight, smell, touch, taste and hearing—as well as how they work are the subjects of qualitative observation. This is about traits or explanations rather than measures or statistics.

9.4 Designs of Qualitative Research

The major research designs in vogue under qualitative research are:

9.4.1 Phenomenological Research

Based on live experiences of participants for the event in which they are involved as described by them, the researchers investigate a phenomenon or event. Discussion, interviews, observation, surveys, etc. can be used to collect information on phenomena. In business, research on the selling method used by sales representatives can be assessed using this method.

Other Examples of Phenomenological Research

- Live experience of tsunami victims.
- Experience of victims of a gas tragedy.
- Vital experience of adults having hearing loss.

9.4.2 Ethnographic Research

The most thorough observational method for studying people in their natural settings is ethnographic research. Here the researchers are required to adjust to the natural settings of the study population, which may be any place from a large city to an organization or can be any isolated place. Here, geographic restrictions may provide a problem for data collection. Understanding the cultures, difficulties, motives and environments that arise can be the goal of this study design. Rather than depending just on conversations and interviews, the researchers actually visit the natural environments. Due to its intensive observational methodology and data collection on those grounds, this sort of study approach might span several days to many months. It is a difficult and time-consuming process that only depends on the researcher's expertise to observe, analyze and infer from the situation. It mostly relates to research for the collection and analysis of data about cultural groups. It is used in studies on the life process of people. Researchers become part of the people who are studied to know their culture by living with them. It is basically to understand people's behaviour in peculiar circumstances.

Example: An ethnographic study on the features, critical attributes, processes and benefits of SHGs of women living with liquor addicted husbands.

9.4.3 Grounded Theory

Researchers collect data to study social processes and social structures and develop theories inductively. It is basically to know the reasons behind actions being taken by people and to develop theoretical models based on existing data in existing modes of genetic, biological or psychological sciences. In business, this method is used for customer satisfaction surveys as to why consumers use the company's product which in turn helps the company to maintain customer satisfaction.

9.4.4 Historical Models

Past events are critically analyzed to understand the present and to anticipate future choices. The experience of the past is used to devise newer methods for more effectiveness in the future.

9.4.5 Narrative Research

In this method, the researchers share stories to understand how participants perceive and make sense of their experiences. It takes subjects to a starting point and reviews situations as problems occur during the course. The business can make use of the results to devise innovations that appeal to the target markets.

9.4.6 Case Study Method

It is a specific research methodology involving a close-up, in-depth and detailed examination of a particular case or cases within a real-world situation though not very common. The case study approach is defined as an intensive study of a single unit or a small number of units (cases) for the purpose of understanding many such groups of similar units. It is an intensive study of an organization, a group, a person or an event which has some specific features compared to many others in the same category. In medical sciences it may be focused on an individual patient or ailment of typical nature, in business it may be focused on a firm with extraordinary performance, in social sciences it may be focused on a social group activity of exemplarity and so on. Case studies can be on more than one unit in the same case study. Such studies are called **cross-case research**.

Over the past several years, the case study approach has matured into a useful, high-quality research technique. It meets the purpose of describing an organization or other entity, as its name implies. Several fields employ this kind of research methodology, including the social sciences and education. Although this approach appears complicated to use, it is one of the most straightforward since it just requires a perfect understanding of the data-gathering techniques and data inference. There are no earmarked designs in the case study method of research. However, the following types of case studies are possible:

- 'No theory first' type case study for methodological work.
- Single case study of rare nature.
- Multiple case studies of similar nature.
- Comparative extreme cases (extreme success and failure cases).

Uses of Case Studies

- Generation of new hypothesis and theories.
- Formulation of new concepts.
- Descriptive richness of cases.
- Explain outcomes of individual cases.
- Add new knowledge to the area of study.

Some Important Points

- Qualitative research relates to the collection and analysis of non-numerical data like verbal or documented information to study concepts, perceptions, experiences, etc. of people.
- Qualitative research focuses on obtaining data through open-ended or conversational communication.
- The strength of qualitative research is its ability to provide complex textual descriptions of how people experience a given research issue.
- Qualitative research methods include in-depth interviews, focus groups, ethnographic research, content analysis, case study research, etc.
- The three most common qualitative methods are participant observation, in-depth interviews and focus groups.
- Qualitative research aims to study social and cultural phenomena based on the live experience of participants and it is an inductive approach to developing new concepts.
- Qualitative research designs emerge during the course of studies and not in advance.
- In person interview is one of the most commonly used techniques in qualitative methods. Only one respondent is interviewed at a time.
- For the focus group, around six to ten respondents from the target population are taken as a focus group. The main aim of the focus group is to find suitable answers to the research questions.
- Record-keeping method uses reliable existing documents as the data source.
- The five sensory organs of seeing, smelling, tasting, hearing and touching are used in qualitative observation.
- Data collection methods in qualitative research include written expression by the participants, observation by the researcher and interactive interviews with the participants by the researcher.
- Ethnographic research is the study of the life process of people. Researchers become part of the people who are studied to know their culture by living with them.

- Grounded theory is basically to know the reasons behind actions being taken by people and to develop theoretical models based on existing data in existing modes of genetic, biological or psychological sciences.
- In the historical model method, past events are critically analyzed to understand the present and to anticipate future choices.
- Case study is a specific research methodology involving a close-up, in-depth and detailed examination of a particular case or cases within the real world.

Suggested Readings

Burns, L. D., and S. J. Lennon, 'Social perception: methods for measuring our perception of others', *International Textile and Apparel Association Special Publication* 5: 153–159, 1993.

Hill, M., 'Research review: participatory research with children', *Child and Family Social Work* 2: 171–183, 1997.

More, J. M., 'Determining sample size', *Qualitative Health Research* 10(1): 3–5, 2000.

Patton, M. Q., *Qualitative evaluation and research methods*, 2nd ed., Sage Publications, Newbury Park, CA, 1990.

Rew, L., 'A theory of taking care of oneself grounded in experience of homeless youth', *Nursing Research* 52(4): 234–241, 2003.

Sampson, P., 'Qualitative research and motivation research', in *Consumer market research handbook*, 3rd ed., edited by R. M. Worcester and J. Downham (eds.), Elsevier, Amsterdam, 1986.

Sharma, B. S., *Research methods in social sciences*, Sterling Publishers Private Ltd., New Delhi, 1983.

Walker, R. (ed.), *Applied qualitative research*, Gower Publishing Company Ltd., Hants, 1985.

Williams, D. G., and N. A. Johnson, *Essentials in qualitative research: a notebook for the field*, McMaster University, Hamilton, 1996.

Yin, R. K., *Case study research: design and methods*, Sage Publications, Beverly Hills, CA, 1984.

10 Design of Sample Surveys

10.1 Sampling Technique

Sampling is a scientific statistical technique widely used when one has to make a decision about a population, universe or aggregate based on the evidence given by a representative sample drawn from that population. Sampling is the method of choosing a part of the population to represent it. It has a wide range of applications in different research methods like experimental, semi-experimental, observational, epidemiological, survey, interview and others in various branches of studies. The decision taken by a doctor on the nature and type of illness of a person based on a test of a few drops of blood taken from the person concerned, or tasting a bit of vegetable being cooked are all day-to-day examples of decisions about a large universe based on a small sample taken from it. Most of the research based on cross-sectional studies makes use of scientific sampling techniques to make decisions about the population/universe under study based on evidence from the sample. Hence sample-based studies form a major approach of research in economics, sociology, education, commerce, agriculture, epidemiology, clinical and other health-related studies. The aim of all sample-based studies is to get measures of basic statistics for the population parameters which are either unknown or need updating. Getting the value of the population parameter is a tedious task due to its large size, widespread nature, cost to collect information from every unit of the population and the time needed to collect and process the data. In such cases, sample-based studies to get estimates of population parameters is a feasible option as the conformity of the sample estimate for population parameters is possible using the theory of probability-based statistical inference. Sample-based survey research is carried out in almost all disciplines, especially in social sciences, commerce, business, management, economics, etc. The use of an appropriate sampling plan, collection of quality data from an appropriate sample size and use of the most appropriate quantitative techniques are the prerequisites for applied quality research.

10.2 Definition of Terms in Sampling

Population: Aggregate of all units on which study is required to be made forms the population.

Examples: Households in a village, students in a school, patients in a hospital, blood cells in a person, insects in a field, soil particles in a field and so on.

Finite Population: A population is finite when there is only a countable number of units in the population.

Examples: Households in a village, students in a school, patients in a hospital.

Infinite Population: A population is infinite when there is an uncountable number of units in the population.

Examples: Blood cells in a person, insects in a field, soil particles in a field, etc.

DOI: 10.4324/9781003527183-10

Target Population: The target population is the group of units on which the study is required.

Examples: All flood-affected people for a study on the impact of floods in a region. All patients with a specific disease for disease-based clinical studies.

Sampled/Accessible Population: Sampled/accessible population is the set of all cases that fulfil the specific standards and are reachable or accessible as subjects for the study by a researcher.

Examples: Flood-effected people living in reachable areas or patients reporting in a hospital.

Sample: A sample is a true representative part of a population under reference so as to make a decision on population parameters based on evidence given by the sample using the theory of probability.

Sampling Unit: The identifiable units of a population on which observations are to be collected.

Sampling Frame: The list of all units in a population from which the sample is to be drawn with assigned probability for each unit of the population to get included in the sample.

Sampling Fraction: The ratio of sample size to population size (n/N) where 'n' is sample size and 'N' is population size is called the sampling fraction.

Population Parameter: The statistical constant like mean (μ), SD (σ), variance (σ^2) and other such measures for a population is called population parameter.

Sample Statistic: The statistical constants like mean ($\bar{X}$), SD, variance (Var) and other such measures calculated using a sample of observations selected from a population is called a sample statistic or estimate.

10.3 Advantages of Sampling

A randomly selected sample, which is a true representative of the population under study, makes it possible to get information about large populations with less cost, less field time and more accuracy. Sampling methods are used when it is not possible to study the whole population due to cost, time or other factors. Sample-based studies have obvious operational convenience over a study of the entire population. It is also possible to make probability-based inferences about population parameters using sample statistics. In other words, generalization of results based on sample studies is possible for randomly selected samples.

10.4 Methods of Sampling

Broadly speaking, the entire sampling method can be grouped as probability samples and non-probability samples. There are different sampling methods under these broad groups which are shown in Figure 10.1.

10.5 Probability Sampling

Probability sampling is also known as random sampling and is a sampling technique in which a sample from a population is selected using the method based on probability theory and every unit in the population has a known probability of getting selected in the sample. Probability sampling makes it possible to estimate population parameters from sample data and confirm their validity through testing of the hypothesis.

10.5.1 Simple Random Sampling

In SRS, every unit in the population has an equal/known probability of being included in the sample. It is applicable when units are more or less homogenous in the population and when

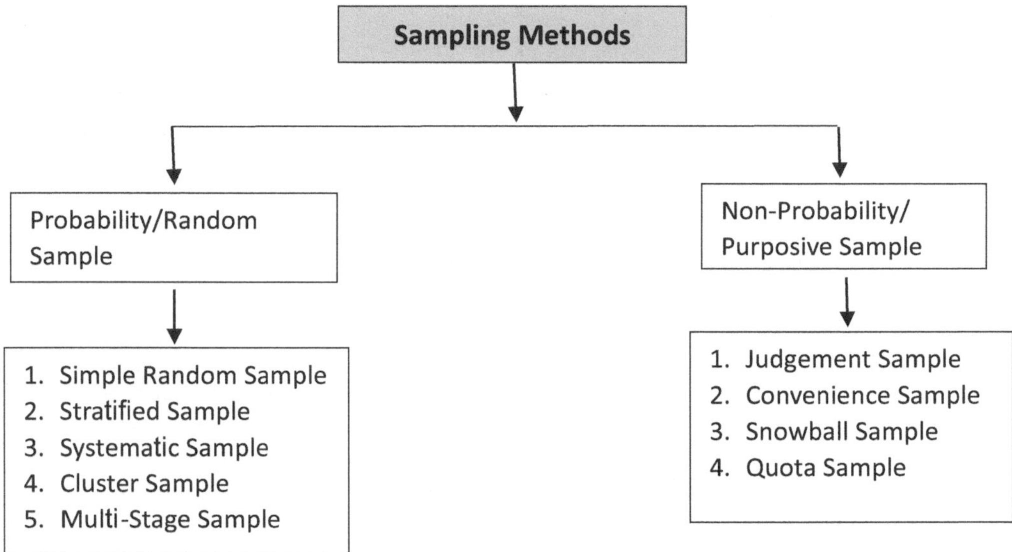

Figure 10.1 Random and Purposive (Non-Random) Sampling Methods

the size of the population is known and a list of units is available. When units with a higher size have a larger probability of being included in the sample, it is called probability proportion to size (PPS) sampling (with or without replacement).

Example: Selection of households from a tribal area for nutritional/family planning adoption/ immunization level of children, etc.

10.5.2 Stratified Sampling

It is the most commonly used sampling plan when the population is heterogeneous and can be grouped into strata/classes which are homogenous within the classes. Random samples are drawn independently from each stratum. Let $N_1, N_2, \ldots N_k$ be the size of strata in the population of size N, then a sample of size $n_1, n_2, \ldots n_k$ drawn from each stratum so that the size of the sample is n. This sampling plan has the advantage of ensuring the representation of each group in the sample and a more precise estimate of population parameters. It has the advantage of generating strata-based and overall estimates of parameters under study. The allocation of sample size in strata can be equal or proportional to the size of strata in the population.

Examples: Students according to classes, patients according to disease, households based on social class, etc.

10.5.3 Systematic Sampling

It is used when some order exists in population units and the population is finite and known in size (serially arranged medical records, patients visiting an OPD). In this plan the first unit is selected at random and subsequent units are selected according to some pre-determined rule.

Let $N = nk$, (N is the size of the population and n is the size of the sample so that $k = N/n$), then one unit is selected at random, say r^{th} unit from first k units in the population. The units at r, $r + k$, $r + 2k$ will form the sample of size n.

Example: Selection of sample patients in a hospital during a certain period when the total number of patients to arrive (N) by the end of the study period is known. If n is the sample, $k = N/n$ is worked out and one number is randomly selected (r) which is known as a random start. Therefore, r^{th} patient arriving for registration will be the first sample unit and subsequent units will be $r + k$, $r + 2k$, from the registration list.

10.5.4 Cluster Sampling

The cluster sampling plan is used when the population under study is widespread and hence the cost of collecting data from selected sample units will be higher. It may be possible to group the final units of study as clusters. In cluster sampling, clusters having a group of final sampling units will be selected randomly and all units in the selected cluster will be included in the final sample. The variability in population will be addressed by selecting a greater number of clusters. The data from all units of selected clusters will be collected and processed.

Example: For a district-level study of a rural health problem based on households, the revenue villages which are well-defined can be treated as a cluster. All the households of selected villages will form the final sample. So, the researcher must randomly select the required number of villages to meet the sample size.

10.5.5 Multi-Stage Sampling

To study a large area (a state or a country) is a tedious job to consolidate a list of all sampling units, to select a random sample and in such cases, multi-stage sampling can be used where ultimate sampling units are selected in stages. To arrive at the ultimate sampling units, the selection process can be done in stages.

For example, for a household-level health survey in a state, the selection of the sample can be made in stages of districts, then blocks, then villages and finally households as ultimate sampling units.

All the sampling plans mentioned above fall under a random sample and have the scope to generalize results after the test of significance.

10.6 Purposive Sampling

The non-random samples, or purposive samples, are generally used to generate quick estimates of unknown parameters which are required in many exigencies, especially in agriculture, health and medical sciences, natural calamities, etc. The application of probability theory is not possible in purposive sampling and hence generalization of results is not feasible. However, the preliminary information and quick results of practical utility can be generated for many purposes.

10.6.1 Snowball Sampling

When a sampling frame is not readily available and as time passes, the size of the sample goes on increasing like a rolling snowball in this type of sampling. It is usually used for studies on populations like sex workers, HIV patients, drug addicts, etc. The available units make it possible to have more new units as the study progresses due to their association with persons of similar type. This method is also called the chain-referral sampling method. This sampling technique can go on and on, just like a snowball, increasing in size (in this case the sample size) till the researcher has enough data to analyze and draw conclusive results.

10.6.2 Quota Sampling

The researchers select a sample according to certain traits or qualities. The population is initially segmented into mutually exclusive subgroups. Then based on judgment, units are selected from different segments of a population when the sample frame is not readily available for the population.

10.6.3 Judgement Sampling

The sample is selected based on the judgement of the investigator. In judgement sampling, the sample units are chosen only on the basis of the researcher's knowledge and judgment. It enables us to select cases that will answer the research question(s) and meet the objectives of the study. Hence it is a purposive sampling method.

10.6.4 Convenience Sample

The sample is selected at the convenience of the investigator in terms of approach or contact.

The application of statistical inference is valid in the case of a random sample. In other words, the generalization of results emerging out of the sample is more valid in the case of a random sample.

10.7 Sample Size

Sample size means the number of samples selected/needed for a study. Sample size determination is the method of selecting the optimum number of sampling units to collect data/observations. An accurate sample size is important to make valid findings from the sample for its generalization for the population under study. The larger the sample size, the more accurate the findings from a study. Generally, if the sample size is 30 or more ($n > 30$), it is considered a large sample. For large samples, the sampling distributions of statistics are normal (Z distribution). If the sample size is less than 30, small sample techniques are used. The sample size determination has a direct bearing on statistical inference-both estimation and testing of hypothesis.

10.7.1 Determination of Sample Size

The basic questions in sampling theory are:

(i) Which sampling technique is to be used for a particular nature and type of population?
(ii) What would be the appropriate sample size to estimate precisely the population parameters?

The answer to the first question can be obtained from the description of the situations explained above. Depending upon the required precision of the estimate to be obtained, the size of a sample for estimating the mean value or a proportion/prevalence can be worked out. A sample larger than what is scientifically needed is a simple waste of time and resources. Similarly, a sample smaller than what is theoretically required will pose problems to the precision of the estimate. Before collecting data, it is important to determine how many samples are needed to perform a reliable analysis. Sample size determination is the statistical assessment of the number of population units to be included in the sample for the study. The sample size must be adequate to represent the population. The determined size should be optimum and must be obtained by the scientific method.

10.7.2 Factors Affecting the Size of the Sample

The sample size for a study depends on many factors. Some of these are:

- **Nature of the Population**: If the population is homogenous, fewer cases will be enough. If the population is heterogeneous, a large size sample will be required.
- **Availability of Resources**: The resources available including time and money are to be considered before determining the size of the sample. Large samples can be taken if sufficient time and money are available.
- **Type of Sampling Method**: If the sampling method is restricted to random sampling, a moderate sample will be enough. In simple random sampling, more numbers may be selected to ensure the representation of all. Cluster sampling demands more samples when compared to stratified sampling.
- **Degree of Accuracy Required**: If a higher degree of accuracy is required, the size of the sample should be large.
- **Nature of Analysis**: Sample size may be influenced by the statistical tools and tests a researcher plans to use for the analysis. Complex multivariate statistical analysis needs larger samples.
- **Factors Used for Sample Size Calculation**: Sample size is influenced by the factors used for sample size calculation such as size of population, margin of error, confidence level, extent of variability, etc.
- **Margin of Error:** The margin of error is a statistic expressing the amount of random sampling error in the results of a survey. The margin of error decreases with an increase in sample size.
- **Confidence Level:** It is the probability that a population parameter lies within a given margin of error with respect to the sample estimate.

Optimum Sample Size: The determined size should be optimum and must be obtained by scientific method. An optimum sample for a study may be defined as that sample which fulfils the requirements of efficiency, representativeness, reliability and flexibility. The sample should be small enough to avoid unnecessary expenses and large enough to reduce sampling errors. The larger size can lead to ethical concerns, time consumption and financial wastage. Smaller sample sizes may cause misrepresentation, inefficiency and insignificant results. An optimum sample size is required to allow for appropriate analysis, to provide the desired level of accuracy and to allow validity of significance test. The common factors considered for the sample size calculations for different research designs are:

- The research hypothesis, null hypothesis (H_0) and alternative hypothesis (H_1) of the research study.
- The two types of errors, Type I error and Type II error and their probabilities as ($\dot{\alpha}$) and (β).
- The precision of the estimate as interval estimate = estimate ± reliability coefficient × SE.
- If study variable X follows N (μ, σ) and $\bar{X}$ is the mean of X, then the confidence interval for,
- $\mu = \bar{X} \pm Z_{(1-\dot{\alpha}/2)}\, \sigma/\sqrt{n}$.
- If study variable X follows N (μ, σ) and p is the estimate of population proportion P and q = 1-p, then the confidence interval for P is given by $= p \pm Z_{(1-\dot{\alpha}/2)}\sqrt{(pq/n)}$.

10.7.3 Methods Determining Sample Size

- **Arbitrary Approach:** It is the rule of thumb method which specifies a fixed percentage of the population as sample size. According to this approach, a sample may be at least 5% of the population and 10% is the ideal.

- **Conventional Approach**: According to this approach, the average sample size used in similar studies can be taken as sample size.
- **Statistical Analysis Requirements Approach**: Here, the sample size is determined based on the proposed statistical analysis considerations. Sample size is determined by the requirements of the proposed statistical techniques for the analysis.
- **Cost-Benefit Basis Method**: In this approach, the sample size is determined based on the availability of resources and the benefits expected. Generally, it is considered ideal for non-probability sampling methods.
- **Confidence Interval Approach**: It applies the concept of variability, sampling distribution and standard error of the estimate. Several formulas have been suggested by statisticians. The elements generally needed to determine the sample size include the amount of variability in the population, desired accuracy (acceptable sampling error), level of confidence or precision level, etc.

Every researcher must identify the primary and secondary outcome variable of the study to state the research hypothesis to assess the required sample size. The outcome variable can be continuous or discrete depending upon it as measurements or counts. The researcher must identify a few published research papers comparable to the research topic under study. The estimated proportion/prevalence, mean and standard deviation of outcome variable are estimated as per the published papers comparable to the topic of study. When the standard deviation of the parameter under study is not available at least the possible range of data (maximum-minimum) as per scientific norms may be used (since range $\approx$ 6 SD).

10.7.3.1 Sample Size for Estimation of Population Mean Values from Cross-Sectional Studies

The width of the interval estimation of a mean is determined as = reliability coefficient × standard error of the estimator. If the reliability coefficient is fixed at 5% or 1%, the width of the interval can be reduced by reducing the standard error. The standard error of the mean is worked out as $\sigma/\sqrt{n}$. When σ is known; the low standard error is possible by increasing the sample size.

If sampling is with replacement from an infinite/large population we can have the following relationship for the width of the confidence interval for the mean.

$$d = \text{reliability coefficient} \times \text{SE of the estimator} = z\alpha \times \frac{\sigma}{\sqrt{n}}$$

$$\text{or } d^2 = \frac{Z\alpha^2\sigma^2}{n} \text{ or } n = \frac{Z\alpha^2\sigma^2}{d^2}$$

Where z_α = Stawwormal Distribution (SND) Variate at α level of significance (5% or 1%), σ = Standard Deviation (SD) of population and d = precision level.

Here the knowledge of σ^2 is required to work out the sample size. Normally it is not available. It can be worked out either by conducting a pilot survey or any available estimate of σ^2 from previous studies can be used or one can use the knowledge of a range of population parameters under the study and the relationship. Range = 6σ hence σ = R/6. The precision level d of the estimate is fixed by the researcher in advance.

10.7.3.2 Sample Size for Estimation of Prevalence of Disease/Population Proportion

$$n = \frac{Z\alpha^2 pq}{d^2}$$

Where, Z_α = standard normal value at α% level of significance (= 1.96 at a = .05 (5% level of significance),

p = prevalence rate of diseases of study or proportion/probability in favour of binary outcome variable (to be obtained from other similar comparable published studies or secondary data sources),

q = 1 − p as p + q = 1; d = precision level of the estimated value of p (width of the confidence interval for population value of P to be decided by the researcher (team).

10.7.3.3 Sample Size for Randomized Controlled Trials (RCT Trials)

The sample size for the RCT-based study is selected by assuming the level of significance or probability of Type I error (a = 0.05 or 0.01); the probability of Type II error (β = 0.20 or β = 0.10), power of test as 1-β; reasonably assessed minimum important difference (MID) of outcome values that would be worth detecting by the researcher; population standard deviation based on other studies of comparable type (σ), carefully defined outcome of interest for the study, equal allocation of study participants (*n*) between treatment and control groups and the statistical null hypothesis of equivalence of outcome variable for the treatment and control groups.

(i) **Sample Size for Outcome Variable Measured as Quantitative Variable**

$$n = \frac{\left(Z_{\alpha/2} + Z_\beta\right)^2 \times 2\sigma^2}{\left(\mu_2 - \mu_1\right)^2}$$

Where $Z_{a/2}$ = normal distribution value at a/2 level of significance;
Z_β = normal distribution value at (1-β) power of the test;
σ^2 = population variance of variable under study which can be taken from other studies or as (Max-Min)/6. In case from any previous studies, $\mu_2 - \mu_1$ = MID (minimum important difference) is available, the same can also be used or the researcher can select the expected MID.

(ii) **Sample Size for Outcome Variable Measured as Qualitative Variable**

$$n = \frac{2\left(Z_{\alpha/2} + Z_\beta\right)^2 p_m q_m}{\left(p_1 - p_2\right)^2}$$

Where $Z_{a/2}$ = normal distribution value at a/2 level of significance;
Z_β = normal distribution value at (1-β) power of test;
p_1 = population proportion of outcome variable/prevalence in the control category;
p_2 = population proportion of outcome variable/prevalence in treatment category.

$$p_m = \frac{p_1 + p_2}{2}; \quad q_1 = 1 - p_1; \quad q_2 = 1 - p_2 \quad \text{and} \quad q_m = 1 - p_m$$

10.7.3.4 Sample Size for Case-Control Studies Based on Qualitative Variable (Cancer as Disease and Smoking as Risk Factor)

In case-control studies, the outcome of cases (group with disease) is compared with control (comparable group with all aspects except disease) for assessing for the risk factor (smoking) for having the disease (cancer). The sample size n is calculated as:

$$\text{Sample Size } n = \frac{r+1}{r}\frac{\left(p^*\right)\left(q^*\right)\left(z_\beta + z_\alpha\right)^2}{\left(p_1 - p_2\right)^2}$$

Where, r = ratio of size in control to size in cases (r =1, when size of case and control is same);

p_1= proportion of exposure to risk factors in cases (from previous studies);

p_2= proportion of exposure to risk factors in control (from previous studies), normally $p_1 > p_2$,

$p*$ = average of proportions exposed in case group and control group, $p^* = \frac{p_1 + p_2}{2}$

$q^* = 1 - p^*$, Z_β = SND variate at the power of the test (0.84 for 80% power and 1.28 for 90 % power) and $Z_{\alpha/2}$ = SND variate at the level of significance at α level (1.96 at 5% and 2.58 at 1% level).

10.7.3.5 Sample Size for Case-Control Studies Based on Quantitative Variable

Suppose that the researcher is interested in studying diabetes in adulthood as a case associated with birth weight (quantitative variable) as a risk factor).

The birth weight is a continuous quantitative variable. The researcher may start with adults with diabetes as a case and non-diabetic as a control. Both groups will be looked back for childhood birth weight.

$$\text{Sample size n} = \frac{(r+1)}{r} \times \frac{\sigma^2\left(z_\beta + z_{\frac{\alpha}{2}}\right)^2}{d^2}$$

Where, r = ratio of size in control to size in cases (r =1, when size of case and control is same),

σ = standard deviation of childhood birth weight (from previous studies),

d = mean difference in case and control of childhood birth weight (from previous studies),

Z_β = SND variate at the power of the test (1-β), that is, (0.84 for 80% power and 1.28 for 90 % power),

$z_{\frac{\alpha}{2}}$ = SND variate at the level of significance at ά level (1.96 at 5% and 2.58 at 1% level).

10.7.3.6 Sample Size for Cohort Studies

In cohort studies, healthy subjects with and without exposure to risk factors are observed. The researcher will start with two groups, one exposed to risk factors and another not exposed to risk factors. Both groups are followed up for having cases either in the past (retrospective cohort) or in the future (prospective cohort). The sample size is calculated as:

$$n = \frac{\left[z_\alpha \sqrt{\left(1+\frac{1}{m}\right)\overline{p}\overline{q}} + z_\beta \sqrt{\frac{p_0(1-p_0)}{m} + p_1(1-p_1)} \right]^2}{(p_0 - p_1)^2}$$

Z_α = SND variate at the level of significance at ά level (1.96 at 5% and 2.58 at 1% level),
Z_β = SND variate at the power of the test (1-β), that is, (0.84 for 80% power and 1.28 for 90% power),
p_0 = probability of case/event in control,
p_1 = probability of case/event in risk factor exposed group,
m = number of control subjects per unit of risk-exposed group.

$$\overline{p} = \frac{p_1 + mp_0}{m+1}$$

$$\overline{q} = 1 - \overline{p}$$

10.7.3.7 Sample Size for Correlation Studies

The sample size for correlation studies to test population correlation coefficient p.

Here null hypothesis H_0: p = 0 and alternative hypothesis H_1: $p \neq 0$.

The level of significance is ά; the power of the test is 1-β and for assumed sample correlation coefficient '*r*'. The sample size is calculated as

$$n = \left(\frac{z_\alpha + z_\beta}{c}\right)^2 + 3$$

where $c = 0.5 \times Ln\frac{1+r}{1-r}$

$Z_ά$ = standard normal variate at ά level of significance,
$Z_β$ = standard normal variate at 1-β level of power of the test.

10.7.3.8 Sample Size for Sensitivity/Specificity of Screening Test (Buderer's Formula)

$$\text{Sample Size for Sensitivity } n = \frac{Z^2_{\left(1-\frac{\alpha}{2}\right)} \times S_N(1-S_N)}{L^2 \times \text{Prevalance}}$$

$$\text{Sample Size for Specificity } n = \frac{Z^2_{\left(1-\frac{\alpha}{2}\right)} \times S_P(1-S_P)}{L^2 \times (1-\text{Prevalance})}$$

n = required sample size for sensitivity/specificity.
SN = anticipated sensitivity.
Sp = anticipated specificity.
1−ά = confidence level (95% or 99%).
L = absolute precision desired on either side half width of confidence interval) of sensitivity/specificity.

Prevalence of the disease under study may be known.
The provision for dropout (10% or 20%) can also be assumed.
The sample size n is calculated for sensitivity and specificity and the higher of the two will be the final sample size.

The sequence of steps in the selection of a sample for any study is depicted in Figure 10.2.

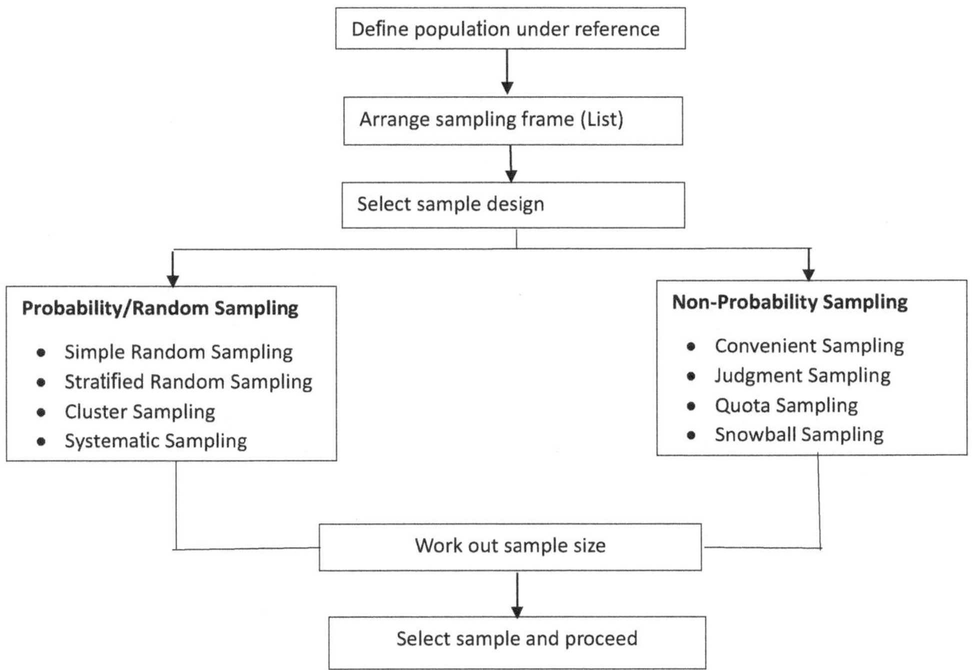

Figure 10.2 Sample Selection Steps

Some Important Points

- **Sampling** is a scientific statistical technique widely used when one has to make a decision about a population or universe or aggregate based on the evidence given by a representative sample of that population.
- **Sample-based studies** aim to get estimates of population parameters which are tested for confirmation.
- **Population parameter**: The statistical constants like mean, SD, variance and other measures calculated for a population are called population parameters.
- **Sample statistic**: The statistical constants like mean, SD, variance and other such measures are calculated using a sample of observations selected from a population is called a sample statistic.
- **Sampling** makes it possible to get information about large populations with less cost, less field time and more accuracy.
- **The sampling methods** can be grouped as probability samples and non-probability samples.

- **In simple random sampling,** every unit in the population has an equal/known probability for being included in the sample, applicable when units are more or less homogenous in the population and when the size of the population is known and a list of units is available.
- **Stratified sampling** is the most commonly used sampling plan when the population is heterogeneous and can be grouped into strata/classes which are homogenous within the classes and random samples are drawn independently from each stratum.
- **Systematic sampling** is used when some order exists in population units and the population is finite and known in size. The first unit is selected at random and subsequent units are selected according to some pre-determined rule.
- **Cluster sampling**, clusters having a group of primary sampling units will be selected randomly and all units in the selected cluster will be included in the final sample.
- **Multistage sampling:** When the area coverage of a study is relatively large, like country or state, for studies at the household level, direct random selection of households is tedious for many reasons. Hence the required sample households will be selected at stages like districts, blocks, villages and then households from selected villages for which the list will be available.
- **For purposive sampling,** random selection criteria are not followed.
- **Snowball sampling:** This purposive sampling method starts with readily available units which makes it possible to have more new units as the study progresses due to their association with persons of a similar type. This method is also called the chain-referral sampling method.
- **Quota sampling:** The population is initially segmented into mutually exclusive subgroups, then based on judgement units are selected from different segments.
- **In judgement sampling**, the sample units are chosen only based on the researcher's knowledge and judgment.
- **Sample size:** For sample-based studies of the population, the sample size must be optimum. Based on the nature of the estimate as mean values or proportion or the study variable as discrete or continuous, the optimum number of samples required must be calculated before the start of the study. The procedure for the selection of samples for different situations is given in Section 10.7.3.

Suggested Readings

Buderer, N. M., 'Statistical methodology I: incorporating the prevalence of disease into the sample size calculation for sensitivity and specificity', *Academic Emergency Medicine* 3(9): 895–900, 1996.

Fox, D. R., 'Computer selection of size-biased samples', *The American Statistician* 43(3): 168–171, 1989.

Golmant, J., 'Correction: computer selection of size-biased samples', *The American Statistician*, 194–194, 1990.

Goodman, L. A., 'Snowball sampling', *Annals of Mathematical Statistics* 32: 148–170, 1961.

Heckathorn, D. D., 'Respondent-driven sampling: a new approach to the study of hidden populations', *Social Problems* 44: 174–199, 1997.

Heckathorn, D. D., 'Respondent-driven sampling II: deriving valid estimates from chain-referral samples of hidden populations, *Social Problems* 49: 11–34, 2002.

Kadam, P., and S. Bhalerao, 'Sample size calculation', *International Journal of Ayurveda Research* 1(1): 55–57, 2010.

Krejcie, R. V., and D. W. Morgan, 'Determining sample size for research activities', *Educational and Psychological Measurements* 30: 607–610, 1970.

Salganik, M. J., and D. D. Heckathorn, 'Sampling and estimation in hidden populations using respondent-driven sampling', *Sociological Methodology* 34: 193–239, 2004.

Singh, D., P. Singh, and P. Kumar, *Handbook on sampling methods*, Indian Agricultural Statistics Research Institute, New Delhi, 1968.

11 Scaling, Coding and Scoring Techniques in Research

11.1 Scale

Scaling Techniques: It is the method of placing respondents in continuation of gradual change in pre-assigned values, symbols or numbers based on the features of a particular object per the defined rules.

All the scaling techniques are based on any or all four aspects i.e. description, order, distance and origin. The research is highly dependent upon the scaling techniques, without which quantitative analysis is not possible in many situations.

A scale is a continuous measurement having a lowest and a highest point, and in between many points. Scaling techniques in research help to quantify qualitative aspects to a quantifiable level. Aspects like attitude, attributes, behaviour, feelings, opinion, etc. can be measured on such scales.

11.1.1 Reliability of a Scale

Reliability of a scale refers to consistency in measurement and can be assessed using any of the following methods:

Re-Test Method: The same scale is applied on all units of a population twice to examine variation, if any, in scaling the objects.

Multiple Form Method: Two types of scales are initially applied to the same group and order of objects in the two scales examined.

Split-Half Method: The scale is divided into two equal parts. Apply both scales to objects of a group. Work out the correlation of scores. A high correlation of scores is treated as reliable.

11.1.2 Validity of Scale

Validity of a scale refers to the accuracy in measurement and can be assessed using any of the following methods:

Logical Validity: The scale must conform to common sense and logic.

Known Group Application: The scale can be applied to known objects and the results compared.

Jury's Opinion: If several jurists give the same scores on the same object, then the scale is valid.

Independent Method: Use two independent criteria to measure the scale value of objects and if results are similar then the scale is valid.

DOI: 10.4324/9781003527183-11

11.1.3 Difficulties in Scaling

The difficulties in scaling include the non-quantifiable nature of qualitative aspects, conceptual problems of qualitative aspects and inconsistent behaviour of the human mind. Additionally, the scales are not universal.

11.1.4 Types of Scale

There are four types of primary scaling systems. These are:

(i) **Nominal Scale:** Nominal scales are adopted for non-quantitative (containing no numerical implication) labelling variables which are unique and different from one another. Types of nominal scales:

Dichotomous: A nominal scale that has only two labels is called 'dichotomous'; for example: yes-no, male-female, rural-urban, etc.

Nominal With Order: The labels on a nominal scale arranged in an ascending or descending order are termed as 'nominal with order'; for example, excellent, good, average, poor, worst.

Nominal Without Order: Such a nominal scale that has no sequence is called 'nominal without order'; for example, black, white, yellow, etc.

(ii) **Ordinal Scale:** The ordinal scale functions on the concept of the relative position of the objects or labels based on the individual's choice or preference. For example, every product has a customer satisfaction level for products according to their buying experience, product features, quality, usage, etc. The ordered ratings so provided are in rank order form like:

Excellent (5); Good (4); Average (3); Poor (2); Worst (1).

(iii) **Interval Scale:** An interval scale is also called a cardinal scale, which is the numerical labelling with the same difference among the consecutive measurement units. The scale has equidistant points, but not having a zero origin. The order and distance between objects can be made out like boxes in a snake and ladder game, thermometer readings, blood pressure readings, etc. With the help of this scaling technique, researchers can obtain a better comparison between the objects.

(iv) **Ratio Scale:** One of the most superior measurement techniques is the ratio scale. It has an interval scale together with a fixed origin (zero level). It can give differences in scores and relative magnitude of scores 15–10 = 25–20. Similar to an interval scale, a ratio scale is an abstract number system. It allows measurement at proper intervals, order, categorization and distance, with an added property of originating from a fixed zero point. Here, the comparison can be made in terms of the acquired ratio.

Table 11.1 will better clarify the characteristics of all four primary scaling techniques including the scope for arithmetic operations and basic statistical calculations.

11.1.5 Other Scaling Techniques

Scaling of objects can be used for a comparative study between two or more objects (products, services, brands, events, etc.).

Table 11.1 Characteristics, Scope for Arithmetic Operation and Basic Statistical Calculation of Primary Scaling Techniques

Particulars/Characteristics	*Nominal Scale (Description)*	*Ordinal Scale (Order)*	*Interval Scale (Distance)*	*Ratio Scale (Description, Order, Distance and Origin)*
Sequential Arrangement	No	Yes	Yes	Yes
Fixed Zero Point	No	No	No	Yes
Multiplication and Division	No	No	No	Yes
Addition and Subtraction	No	No	Yes	Yes
Difference in Variables	No	No	Yes	Yes
Mean	No	No	Yes	Yes
Median	No	Yes	Yes	Yes
Mode	Yes	Yes	Yes	Yes
Range	No	Yes	Yes	Yes
Standard Deviation	No	No	Yes	Yes

Following are the two categories under which other scaling techniques are used based on their comparability:

- **Comparative Scales:** For comparing two or more objects of a study, a comparative scale can be used by the respondents. Following are the different types of comparative scaling techniques:
  - (i) **Paired Comparison:** It is used to select any one out of the two objects/products by the respondents. This technique is mainly used for product testing, to facilitate the consumers with a comparative analysis of the two major products (say, P and Q) in the market. To compare more than two objects say comparing P, Q and R, one can first compare P with Q and then the superior one (i.e. one with a higher percentage) with R.
  - (ii) **Rank Order Comparison Scale:** In rank order scaling the respondent needs to rank or arrange the given set of objects according to his or her preference.

    For example, a soap manufacturing company conducted a rank order scaling to find out the ordering preferences of consumers. The respondents were asked to rank four soap brands in the sequence of their choice as in Table 11.2.

    The above scaling shows that soap 'Y' brand is the first and most preferred brand, followed by soap 'X', then soap 'Z' and the least preferred one is soap 'V'.

Table 11.2 Response of Respondents on their Preference of Four Soap Brands

SOAP BRANDS	*RANK*
V	4
X	2
Y	1
Z	3

(iii) **Constant Sum Comparison Scales:** Is a scaling technique where a continual sum of units is allocated to the given features, attributes and importance of a particular product or service by the respondents.

For example, three respondents belonging to three different social groups are asked to allocate 100 points to the following attributes of a soap product 'S'. Their response is given in Table 11.3.

Table 11.3 Response of Three Respondents on Features of New Soap 'S' Using Constant Sum Scale

Attributes	*Respondent 1*	*Respondent 2*	*Respondent 3*
Appearance	22	16	18
Skin Protection	22	24	16
Smell	14	22	24
Packaging	18	16	20
Unit Price	24	22	22
Total	**100**	**100**	**100**

From the above constant sum scaling analysis, we can see that:

Respondent 1 considers product 'S' due to its competitive price as a major factor.
Respondent 2 preferred the product because it is skin protection.
Respondent 3 preferred the product because of its smell.

In such situations, constant sum scales can be used to identify the most preferred feature out of all given features of the object, product or services.

(iv) **Q-Sort Scaling:** Q-sort scaling is a technique for sorting into groups of uniform nature or preference out of a given large number of objects. It emphasizes the ranking of the given objects in descending order to form similar piles based on specific attributes. It is a modified form of rank-order scaling.

For example, the marketing manager of a garment manufacturing company sorts the most efficient marketing executives based on their past performance, sales revenue generation, dedication and growth. The Q-sort scaling can be performed on a group of executives, and the marketing head can create a few groups based on their efficiency.

- **Non-Comparative Scales:** A non-comparative scale is used to analyze the performance or choice of an individual product or object on different parameters. Following are some of the most common types:

(i) **Continuous Rating Scales**

It is a graphical rating scale where the respondents are free to place the object in a position of their choice. It is done by selecting and marking a point along the vertical or horizontal line which ranges between two extreme points.

For example, a mattress manufacturing company used a continuous rating scale to find out the level of customer satisfaction of its new comfy bedding. The response can be taken in the following ways:

Rating Scale: It gives position to individuals. There are two types of rating scales. (i) Graphic rating, (ii) itemized rating.

- **Graphic Rating**: Rater indicates his rating position by ticking on a graphic scale such as excellent (1), very good (2), good (3), average (4), below average (5), poor (6), very poor (7).
- **Itemized Rating**: The rater selects one of the limited choices which are given in terms of scale position on a 0–5 or 0–10 scale.

(i) **Itemized Rating Scale:**

Itemized rating scale is a widely used technique under the non-comparative scales. It emphasizes on choosing a particular category among the various given categories by the respondents. Each class is briefly defined by the researchers to facilitate such selection.

The three most commonly used itemized rating scales are as follows:

(ii) **Summated Scale (Likert Scale):**

The Likert scale was developed by Rensis Likert, a psychologist in 1932. A definite number of favourable and unfavourable statements are used. The respondent has to react to each of these statements and express the degree of agreement or disagreement. Each of his/her responses is given a numerical score in the order of degree. The total score for each respondent is worked out by adding scores of each statement. In the Likert scale, the degree of agreement is given scores as in Table 11.4.

Table 11.4 Scores Given on a Likert Scale on the Basis of Degree of Agreement

Degree of Agreement	*Scores*		
Strongly approve	05	or	+2
Approve	04	or	+1
Undecided	03	or	0
Disapprove	02	or	−1
Strongly disapprove	01	or	−2

The Likert scale makes it possible to transform ordered qualitative aspects into quantitative mode so that quantitative techniques applicable to numerical data can be used to analyze data for further inferences.

In the Likert scale, the researcher provides some statements and asks the respondents to mark their level of agreement/disagreement or satisfaction/dissatisfaction over these statements by selecting any one of the given options from the alternatives.

For example, a shoe manufacturing company adopted the Likert scale technique for its new sports shoe range named Z sports shoes. The purpose is to know the agreement or disagreement of the respondents.

For this, the researcher asked the respondents to tick any one option representing the most suitable answer according to them, out of the following options:

- Strongly Disagree (SD)
- Disagree (D)
- Neither Agree nor Disagree (NAND)
- Agree (A)
- Strongly Agree (SA)

The illustration in Table 11.5 will help the company to understand what the customers think about its products. Also, whether there is any need for improvement or not.

(iii) **Semantic Differential Scale:**

It is a scale used for attitudinal assessment. Attitudes of people on objects, concepts, ideas, etc. are assessed using this scale. It consists of bipolar adjective pairs of extreme situations like beneficial-harmful, good-bad, wise-foolish, etc. Hence it is a type of

Table 11.5 Use of Likert Scale to Get the Degree of Agreement for a New Sports Shoe Range

Feature	*SD*	*D*	*NAND*	*A*	*SA*
Very lightweight					
Durable					
Cost-effective					
Extremely comfortable					
Look too trendy					
Recommend it to others					

differential scale used to derive the respondent's attitude on certain objects, ideas, events, etc. It is commonly used for purposes like customer satisfaction, employee satisfaction, etc. Depending upon the nature of the commodity, idea, concept, object, etc., the commonly used bipolar adjectives are good-bad, hard-soft, fast-slow, hot-cold, difficult-easy, happy-sad, etc. The semantic differential scale on attitude toward a new product by customers can be assessed by selecting any logical value between k to -k for aspects like taste (good to bad), colour (bright to dull), keeping quality (long to short), etc.

In a bipolar seven-point non-comparative rating scale the respondent marks any of the seven points for each given attribute of the object as per personal choice, thus depicting the respondent's attitude or perception towards the object.

For example: A well-known brand for watches carried out semantic differential scaling to understand the customer's attitude towards its product. The customers are required to tick the most appropriate option for each of the attributes related to the watch. The representation of this technique is given in Table 11.6.

Table 11.6 Semantic Differential Scale

From	*+3*	*+2*	*+1*	*0*	*-1*	*-2*	*-3*	*To*
Stylish	.....	....	.....	.....	.....	.....	....	Unfashionable
Simple	......	.....	......	......	....	....	....	Complex
Affordable	......	....	......	......	....	....	......	Expensive
High Quality	....	....	....	....	....	....	....	Low Quality
Wide Variety	....	....	....	....	....	....	....	Limited Variety

From the above diagram, one can analyze customers' preferences and non-preferences for the given feature of the product. Where more ticks on negative points are observed, the company can make modifications to the product to make it more appealing or positive for such features.

(iv) **Visual Analogous Scale (VAS):**

The VAS scale is used to rate attributes/feelings/sensations like pain, shivering, cold, etc. on a 10-point line depending upon intensity. The line can be horizontal or vertical. VAS is a 10 cm or 100 mm line with extreme points at both ends like no pain to severe pain, worst quality to best quality, highly alert to extreme dullness and so on. The VAS scales are used in medical sciences to assess the magnitude of pain while undergoing treatment or intervention as part of research work.

(v) **Stapel Scale:**

A Stapel scale is an itemized rating scale that measures the response, perception or attitude of respondents toward a particular object through a unipolar rating. The range of a Stapel scale is between −5 to +5 eliminating zero, thus confining it to ten units.

For example, A tour and travel company asked the respondent to rank their holiday package in terms of value for money and user-friendly interface as in Table 11.7.

Table 11.7 Scale Used by Tour and Travel Company to Know Customer Satisfaction

Value for Money	*User-Friendly Interface*
+5	+5
+4	+4
+3	+3
+2	+2
+1	+1
0	0
−1	−1
−2	−2
−3	−3
−4	−4
−5	−5

If a greater number of respondents tick the positive values of the upper range for a particular aspect, it shows higher customer satisfaction. On the other hand, if more respondents tick the negative values radical changes are to be made for those aspects and low-value positive ticks also, special attention is needed to improve the choices of the people/respondents.

11.1.6 Other Medical Scales

There are a large number of scales and scores developed by medical scientists for specific purposes, some of these are given in Table 11.8.

Table 11.8 Some of the Scales/Scores Used in Medical Research

S. No.	*Name of scale/score*	*Developed by*	*Year of development*	*Purpose*
1	Apgar	Virginia Apgar	1952	To quickly evaluate health of newborn children
2	ASA physical status	American Society of Anesthesiologist	1963	For assessing fitness of patient before surgery
3	Asthma Life Impact Scale (ALIS)	Galen Research	2010	To measure asthma's effect on quality of life
4	Athens Insomnia Scale	Research from Greece	2000	To assess the insomnia symptoms in patients with sleep disorders
5	Ballard Assessment	Jeanne L. Ballard	1979	For assessment of gestational age
6	Barthel Scale	Mahoney, F.I. and Barthel, D.W.	1965	To measure the performance of activities in daily life
7	Beck Hopelessness Scale (BHS)	Aaron T. Beck		To measure hopelessness feeling about future, loss of motivation and expectations
8	Blantyre Coma Scale	Terrie Tylor and Malcolm Molyneux	1987	To assess malarial coma in children

(*Continued*)

Table 11.8 (Continued)

S. No.	*Name of scale/score*	*Developed by*	*Year of development*	*Purpose*
9	Bristol Stool Scale	Bristol Royal Infirmary	1997	To classify type of feces (diagnostic triad for irritable bowel syndrome)
10	Disability Rating Scale (DRS)	Mrappaport	1982	To track traumatic brain injury
11	Glasgow Coma Scale (GCS)	Bryan Jennett and Graham Teasdale	1960	To assess persons level of consciousness after brain injury
12	Injury Severity Score	Association for the Advancement of Automotive Medicine	-	To assess mortality, morbidity and hospitalization time after trauma
13	Malinas Score	Yves Malinas	1997	To determine the birth time of pregnant woman
14	Morse Fall Scale (MFS)	Morse J.M., Springer	2008	To assess a person risk of fall
15	Perceived Stress Scale	Sheldon Cohen	1983	To measure extent of stress in the life
16	Pain Assessment in Advanced Dementia (PAINAD)	Victoria Warden, et al.	-	To assess pain experienced by those with dementia
17	Scoring Atopic Dermatitis (SCORAD)	European Task Force on Atopic Dermatitis	1993	To assess severity of atopic dermatitis
18	Tanner Scale	James Tanner	1969	To measure the maturity level of the children

11.1.7 Errors in Scale

We know that the errors in sampling techniques include sampling and non-sampling sources. The sampling errors can be minimized by increasing the sample size whereas non sampling errors are to be minimized through training and supervision of data collection personnel and by using error free equipment to record such data. The measurement errors fall under non sampling errors. The errors in scale are a part of non-sampling error which includes:

- Fault of instruments—weak battery.
- Respondent's indifference—non concern.
- Technician's lack of knowledge/in-experience.
- Situational error like recording blood pressure of a patient just after climbing staircases.

11.2 Coding

Coding is a technique for scientific communication between the human mind and computer system. It is transformation of human language to computer language. Coding is a process for identification and classification by the computer. Quite often, the data collected for research purposes include both numerical and non-numeric information like names, attributes, etc. The nonnumeric data of such types will have to be coded for the purpose of computer manipulations.

Coding can be numeric (01, 02 or alphanumeric A-01, A-02). In the International Classification of Diseases (ICD) all diseases are coded with alphanumeric codes having four characters as per WHO guidelines. In the Farm Analysis Package (FARMAP) of FAO, all agricultural commodities, inputs, output, etc. are numerically coded according to record type. For entering research data in MS EXCEL, researchers can develop codes for descriptive data and information for items like name of respondent, location, sex, etc. along with other qualitative information.

11.3 Scoring

These days scorecards are used for appointment, promotion, etc. of people based on academic merit, experiences, etc. to make the process more objective-oriented and transparent. Scoring is a method of generating numerical data as scores for research variables from the respondent's responses on multiple choice options, dichotomous or qualitative responses, etc. KAP (knowledge, attitude and practice) study is a research method widely used in agricultural extension, community health, education, etc. For each of the selected respondents of the study, scores are generated for knowledge, attitude and practices using objective-type questions. Once the scores are generated for knowledge, attitude and practice based on objective statements for each of the respondents, then statistical analysis as applicable to qualitative data can be done. Sometimes the aggregate of scale values of different aspects of an item is treated as a score. In KAP studies the scores for knowledge, attitude and practice can be developed for studies like the KAP study on farmers for organic farming, KAP study on family planning adoption by adults of reproductive age, KAP study on prevention of COVID (mask, hand washing and social distancing) in rural areas, etc.

Global Assessment Functioning (GAF) Score: GAF scoring system is that mental professionals use to assess the effective functioning of the daily life by a person. It is also used to measure the impact of psychiatric illness on the personal life, skills and abilities of a person. The score ranges from zero to 100. It is based on the level of functioning in daily life and the severity of mental illness of a person. Doctors determine GAF score based on conversion with the patient, their family members or caretakers, review of medical records and on examining their behavioral history including police/court records. Though the score is numerical, it is more subjective. As the score declines, the severity of mental illness increases.

Some Important Points

- A scale is a continuous measurement having number of points from a minimum to a maximum.
- Reliability of a scale refers to consistency in measurement.
- Validity of a scale refers to the accuracy in measurement.
- Nominal scales are adopted for non-quantitative labeling variables which are unique and different from one another.
- Ordinal scale functions on the concept of the relative position of the objects.
- Interval scale is also called a cardinal scale which is the numerical labeling with the same difference among the consecutive measurement units. The scale has equidistant points, but not having a zero origin.
- Ratio Scale has an interval scale together with a fixed origin (zero value).
- Under comparative scaling technique paired comparison is made by two objects at a time, rank order comparison is made by giving ranks to each of the objects.
- Under the constant sum method of scaling, normally used in product differentiation, the attributes of the product are given marks by the respondents in such a way that the total marks given by each respondent is a constant number fixed in advance.

- Q-sort scaling is a technique used for sorting the most appropriate objects out of a large number of given variables. It emphasizes the ranking of the given objects in a descending order to form similar piles based on specific attributes.
- A non-comparative scale is used to analyze the performance of an individual product or object on different parameters.
- Rating Scale gives position to individual objects on an overall basis. There are two types of rating scales. (i) Graphing rating, (ii) itemized rating.
- Under graphic rating the rater indicates his rating position by ticking on a graphic scale such as excellent (1), very good (2), good (3), average (4), below average (5), poor (6), very poor (7).
- Under itemized rating the rater selects one of the limited choices which are given in terms of scale position on a 0–5 or 0–10 scale.
- Under the summated scale (Likert Scale), a definite number of favourable and unfavourable statements is used, and the respondent has to react to each of these statements and express the degree of agreement or disagreement.
- Semantic differential scale is used to express the attitude of people on objects, concepts, ideas, etc. based on bipolar adjective pairs of extreme situations like beneficial-harmful, good-bad, wise-foolish, etc.
- Visual analogous scale (VAS) is used to rate attributes/feelings/sensations like pain, shivering, cold, etc. on a ten-point line depending upon intensity.
- A Stapel scale is that itemized rating scale which measures the response, perception or attitude of the respondents toward a particular object through a unipolar rating. The range of a Stapel scale is between −5 to +5 eliminating zero, thus confining to ten units.
- Medical scientists have developed a large number of scales to measure qualitative medical parameters in research.
- Coding is a technique for scientific communication between the human mind and computer system.
- Scoring is a method of generating numerical data as scores for research variables out of respondents' responses on multiple choice options, dichotomous or qualitative responses, etc.

Suggested Readings

Bentler, P. M., and D. G. Weeks., 'Restricted multidimensional scaling models', *Journal of Mathematical Psychology* 17: 138–151, 1978.

Blalock, H. M., *Social statistics*, McGraw-Hill, New York, 1972.

Commandeur, J. J. F., and W. J. Heiser, *Mathematical derivations in the proximity scaling (PROXSCAL) of symmetric data matrices*, Department of Data Theory, University of Leiden, Leiden, 1993.

Green, P. E., and V. Rao, *Applied multidimensional scaling*, Dryden Press, Hinsdale, IL, 1972.

Jones, L. E., and F. W. Young, 'Structure of a social environment: longitudinal individual differences scaling of an intact group', *Journal of Personality and Social Psychology* 24: 108–121, 1972.

Kristof, W., 'Estimation of true score and error variance for tests under various equivalence assumptions', *Psychometrika* 34(4): 489–507, 1969.

Nishisato, S., *Analysis of categorical data: dual scaling and its applications*, University of Toronto Press, Toronto, 1980.

Schiffman, S. S., M. L. Reynolds, and F. W. Young, *Introduction to multidimensional scaling: theory, methods and applications*, Academic Press, New York, 1981.

Torgerson, W. S., *Theory and methods of scaling*, Wiley, New York, 1958.

12 Research Variables and Research Data

12.1 Research Variables

The research methods and designs in quantitative research aim to generate research data or evidence for the research variables applicable to a specific topic of research. Hence the knowledge of broad classes of research variables and research data is important, especially in the context of the application of statistical software like SPSS, SAS, R-programming, etc.

The broad classes of research variables and their meanings are:

Independent variable: It is that category of variables representing a specific and specified item of measurement in research which other variables in the study cannot alter, but it can change/influence the values of many other variables. For example, age, height, weight, etc. in health studies; rainfall, inputs, etc. in agricultural production studies. In regression analysis of type Y = F(X), X is the independent variable.

Dependent variable: It depends on the values of other variables. Normally, independent variables can change the values of dependent variables. For example, customer satisfaction as a dependent variable (Y) depends on customer services as an independent variable (X), or crop production (Y) as a function of the independent variable fertilizer (X). In regression analysis, Y = F(X), Y is the dependent variable.

Explanatory variable: Independent variables in regression analysis are also called explanatory variables (X). Values of these variables can explain the variations in values of the dependent variable (Y). In cause-and-effect studies, the changes in the values of these variables (X) cause a corresponding change in the values of the dependent variable (Y). If $Y = F(X_{1,} X_2, \ldots, X_k)$, variables $X_1, X_2, \ldots, X_k$ are explanatory variables.

Dummy/Proxy variable: In many studies, the explanatory variables can be qualitative variables like sex, religion, season, etc. which cannot be measured in numerical form, but can influence the dependent variable of the study. A dummy variable is a constructed variable to describe variation in the outcome/dependent variable due to such qualitative variables of the study. Normally such variables are included as explanatory/independent variables in regression analysis. When quantitative data like exact age is not available, age groups such as young or old can be included as dummy explanatory variables.

Lagged variables: In regression analysis like acreage response functions of crops, consumption function of families, etc. the current period values may depend on lagged (previous period) values of the dependent variable itself as well as lagged and current values of other explanatory variables. For example, the consumption function may be of the form $C_t = F(C_{t-1}, C_{t-2}, Y_t, Y_{t-1}, X_{1t}, X_{2t}, \ldots)$. Where C_t is the tth year consumption (dependent variable) and independent/explanatory variables include one year lagged consumption (C_{t-1}), two year lagged

DOI: 10.4324/9781003527183-12

consumption (C_{t-2}), current year income (Y_t), one year lagged income (Y_{t-1}), current year family size (X_{1t}), current year livestock population (X_{2t}), etc.

Outcome variable: Most of the research, especially experimental research, is focused on an outcome variable which depends on the values of independent variables/interventional variables. The academic performance/marks obtained by a student in an examination depend on his/her attendance in class or daily hours of study at home/hostel. Here, marks obtained are the outcome variable of a study on academic performance. Generally, the outcome variable is the dependent variable of the study. In medical research, the onset time and duration of sensor or motor block as a response to anaesthesia drugs are the outcome variables. Similarly, the healing time in surgical methods or duration of discharge from hospital etc. are outcome variables.

Endogenous variable: These are variables in a statistical model that are changed or determined by their relationship with other variables; therefore endogenous variables are dependent variables. In demand-supply studies, price is an endogenous variable as supply can cause changes in prices.

Exogenous variable: These are variables not included in the system but influencing the outcome variable from outside like environmental factors like atmospheric temperature, hailstorm, wind, etc. in an input-output study of agricultural production.

Intervening variable: It is a mediator variable that links the dependent and independent variables in a study. For example, rainfall or sun intensity are intervening variables in a study on fertilizer response to agricultural production.

Control variables: Control variables are those kept constant or fixed during the study. For example, in the agricultural production study, the crop variety was kept the same for all the fields as a control variable.

Moderating variables: A moderating variable can influence the relationship between the dependent and independent variables through its presence. In a study on agricultural production, the climatic factors work as a moderating variable.

Extraneous variables: Extraneous variables are those factors the researcher failed to consider while planning an experiment. For example, the basic fertility level of the soil in a fertilizer response study which the researcher has missed while planning the experiment is an extraneous variable. It can influence the outcome variable and hence lead to erroneous conclusions.

Quantitative/Metric variables: Quantitative variables are those that can get values as numbers of any type. For example, family size, family income, family expenditure, etc., in household economy-related studies.

Discrete Quantitative variable: It includes those quantitative variables taking values as counts. For example, family size, number of children, number of dependent family members, number of milk animals, etc. in rural household-based studies.

Continuous Quantitative variable: It includes those variables taking values as measurement, height, weight, age, etc. For example, income and consumption expenditure, income expenditure-based studies and blood pressure or sugar level in health studies.

Qualitative/Non-Metric/Categorical variable: It includes non-numerical values or data on a nominal scale. For example, male or female; joint family or nucleolus family; rural or urban family, etc. in household-based studies.

Binary/Dichotomous variable: It includes variables that can fall only in any one of the two possible categories. For example, the sex of newborn child as male or female, exam result as pass or fail, dietary type as vegetarian or non-vegetarian, place of stay as rural or urban, etc.

Nominal variable: A nominal variable is a type of variable that is used to name, label, or categorize an attribute as a part of a research study. It includes qualitative variables that can fall into categories without any order. For example, patients as male or female in health research, literacy level of the person as literate or illiterate, etc.

Ordinal variable: An ordinal variable is a categorical variable with ordered values. It is a variable that is in between categorical and quantitative variables. It includes variables that can fall into three or more categories in some order. For example, level of satisfaction, level of pain, etc. expressed in Likert and other scales.

Confounding variable: A confounding variable is an unmeasured third variable in a study aimed at cause-and-effect relationships. It includes those variables that are not included in the study but can disguise the effect of another variable. Such variables can distort the results by making them biased. For example, in a study on the consumption of a commodity with the income of the consumer, the climatic factors act as a confounding variable.

Composite variable: The composite variables are generally categorical in nature. When two or more variables are combined to form a new variable, it becomes a composite variable. For example, body mass index (BMI) is a composite variable of height and weight of a person as BMI = weight in kg ÷ height in meters.

12.2 Research Data

Data is the quantitative or qualitative identity of various physical, chemical and biological entities which are either measured or observed. Scales of measurements provide nature and magnitude to data with respect to any study unit or respondent of the study. The four basic scales of measurement are nominal scale, ordinal scale, interval scale and ratio scale. The nature, type and size of data collected for any research largely depend upon the aims and objectives included in the plan of the research study. The numerical data plays a crucial role in applied research. Research data can be considered as the values of quantitative or qualitative variables included in the study. Data without processing is called raw data and after processing it becomes useful information. The following are broad categories of data:

Primary data: Data directly collected the first time from a source or unit of study are primary data. Most of the research data generated through experimental, survey and observation-based research are primary data. The applied research taken by research scholars is mostly primary data-based.

Secondary data: The data collected for certain specified purposes or as mandatory requirements by official and non-official agencies and documented is called secondary data. Most of the official data are available on a time series basis across units and are rich sources of data for secondary data-based research to assess growth, trends, business cycles, etc.

Cross-Sectional data: Data collected for many subjects/units of study from a specified space at one point in time is cross-sectional data. Most of the data generated using experimental, survey and observation-based methods fall under cross-sectional data. A large number of quantitative techniques are available for analysis of cross-sectional data which depends on the aim and objectives of the study.

Categorical data: Qualitative data which cannot be measured numerically is called categorical data. Such data are also called attributes. For example, race, sex, hair colour, educational level, etc.

Univariate data: Data/observation collected on a specific characteristic. It can be numeric or categorical. For example, farmer-wise production of a crop.

Time Series data: The data related to a variable according to time periods is called time series data. For example, the decadal population of India from 1951 to 2011 or the daily temperature of a patient.

Spatial data: It includes geospatial data or geographic data. It is used in geographic information systems (GIS).

Ordered Data: It includes -data in an ordered type like low, medium, high.

12.2.1 Broad Types of Research Data

The broad categories of research data are quantitative data and qualitative data. The qualitative data are non-metric, categorical or nominal in nature. The broad categories of quantitative data are:

(i) Discrete data (countable) and continuous data (measurable).
(ii) Primary data (collected for the first time by the researcher) and secondary data (collected by some agencies and used by the researcher).
(iii) Time series data (chronological data of a variable) and cross-sectional data (data for many subjects at one point in time).

12.2.2 Methods of Research Data Collection

Most of the research data are either measured or observed or collected as responses from the subject of study. The following methods of data collection for research have been discussed in previous chapters:

(i) **Quantitative Methods**

- Experimental/interventional method (conduct experiments based on specific design befitting the problem).
- Non-interventional observation method (data collected by the researcher or his/her representatives from subjects of study like OPD and laboratory test data without having any intervention).
- Interview schedule (to be filled by the researcher/investigator by interacting with the respondent) or questionnaire (to be filled by the respondent).
- Online method of data collection (Google Form).

(ii) **Qualitative Methods**

The broad methods of data collection in qualitative research include:

- One-to-one interview.
- Focus group discussion,
- Records and documents.
- Process of observation.

In the previous chapters, we discussed the methods of generating research data. In the succeeding chapters, we will discuss the methods of analyzing research data using some of the basic statistical techniques irrespective of the disciplines of study. In fact, every researcher must be aware of a cafeteria of quantitative techniques so that he or she can select the most appropriate technique for a given problem. The data collection format for any research study must take into account the data requirement to address the stated aims and objectives of the study. Besides, the researcher must have a plan for data analysis while preparing the research synopsis or plan of research.

Some Important Points

- Independent variables are specific and specified items of measurement in research which other variables in the study cannot alter but can change/influence the values of other variables. Independent variables are also called explanatory variables.
- Dependent variable depends on values of other variables, that is, independent variables.
- Outcome variables are the focal variable of research which depends on the values of independent variables of the study.
- Endogenous variables are variables causing changes by staying within the system.
- Exogenous variables are variables not included in the system but influencing the outcome variable from outside.
- Control variables are those kept constant or fixed during the study.
- Extraneous variables are those factors that the researcher has failed to consider while planning an experiment.
- Quantitative/metric variables are those which can get values as numbers of any type.
- Discrete quantitative variables include those quantitative variables which take values as counts.
- Continuous quantitative variables include those variables taking values as measurement.
- Qualitative/non-metric/categorical variables include non-numerical values or data on a nominal scale.
- Nominal variables are used to name, label or categorize an attribute as a part of a research study.
- An ordinal variable is a categorical variable with ordered values.
- A confounding variable is an unmeasured third variable in a study aimed at cause-and-effect relationships.
- Data directly collected for the first time from a source are primary data.
- The data collected for certain specified purposes or as mandatory requirements by official and non-official agencies and documented for use by researchers are called secondary data.
- Data collected for many subjects from a specified space at one point of time is cross-sectional data.
- The broad categories of research data include discrete data (countable) and continuous data (measurable); primary data (collected for the first time by the researcher) and secondary data (collected by agencies and used by the researcher) and time series data (chronological data) and cross-sectional data (data for many subjects at one point of time).
- The methods of quantitative research data collection include the experimental method, observation method (in experiment-based research designs the required data is collected by the researcher or his/her representatives), interview schedule (to be filled by the researcher/investigator by interacting with respondent), questionnaire (to be filled by the respondent against each question set) and online method of data collection.
- The methods of qualitative research data collection include one-to-one interviews, focus groups, record keeping and process of observation.
- The data collection format for any research study must take into account the data requirement to address the stated aims and objectives of the study. Besides, the researcher must have a plan for data analysis while preparing the research synopsis or plan of research.

Suggested Readings

Bartholomew, D., M. Knott, and I. Moustaki, *Latent variable models and factor analysis: a unified approach*, John Wiley & Sons, New York, 2011, https://doi.org/10.1002/9781119970583

Heiser, W. J., *Unfolding analysis of proximity data*, Department of Data Theory, University of Leiden, Leiden, 1981.

Hoaglin, D. C., F. Mosteller, and J. W. Tukey, *Exploring data tables, trends, and shapes*, John Wiley & Sons, New York, 1985.

Kass, G., 'An exploratory technique for investigating large quantities of categorical data', *Applied Statistics* 29(2): 119–127, 1980.

Kinderman, A. J., and J. G. Ramage, 'Computer generation of normal random variables (Correction: 85: 212)', *Journal of the American Statistical Association* 71: 893–896, 1976.

13 Basic Statistical Methods for Research

13.1 Introduction

Applied statistics plays a big role at all stages of research and its role is all the more important at the stage of data analysis. Applied research, in different areas, invariably uses basic statistical methods to describe data related to the situation/problem under study or to relate data for different variables of study. Some of these concepts are briefly discussed in the forthcoming sessions:

- Measures of central tendency (arithmetic mean, geometric mean, harmonic mean, median, mode, quartiles, deciles, percentiles).
- Measures of dispersion (range, mean deviation, variance, standard deviation, interquartile range, coefficient of variation (CV)).
- Skewness.
- Kurtosis.
- Probability and basic probability distributions.

These descriptive statistical measures along with frequency tables are widely used in descriptive research methods to explain the research problem with facts and figures.

13.2 Measures of Central Tendency

Every set of data can have various measures of central tendency which is a central representative value for the whole dataset. An ideal measure should be conceptually conceivable and easy to calculate using all of the possible values in the dataset.

Arithmetic Mean (AM)

- All AM, simple mean and average are the same and represent the central tendency of data.
- Applicable for all types of quantitative research methods (experimental, observational, survey, etc.).
- AM of the set of data $x_1, x_2, \ldots . x_n$ is denoted as $= \frac{\text{Sum of all values}}{\text{n}} = \frac{\sum x_i}{n}$.
- More relevant when low or moderate variability in the data, that is, low SE.
- Not a robust measure when outliers and extreme values are available in the data set.

Median

- The median is the middle observation for a dataset sorted in ascending or descending order. The median divides the data into a higher half and a lower half and hence is a positional average.

DOI: 10.4324/9781003527183-13

- To calculate median, arrange the data in ascending or descending order. Then, in the case of odd number of data, the middle value is the median and for even number of data, average of the two middle values is the median.
- When there are extreme values in the dataset, median is the best alternative measure of central tendency.
- In frequency distributions, the class having a median value is called the median class.

Mode

- Mode is the most repeated value in a dataset.
- The probability density function will have maximum probability at mode.
- There can be more than one mode in a set of data.
- For a symmetric unimodel distribution like normal distribution, the mean, median and mode will be the same.
- In a business dataset, mode is important for many decision-making purposes.
- In frequency distributions the class having the highest frequency is called the model class.
- Mode has a wide-range of practical applications in business, economics, education, health sciences, etc.

Geometric Mean (GM)

- Geometric mean of a set of 'n' data is the nth root of the product of all the data.
- GM of $x_1, x_2, \ldots, x_n$ is GM = $\sqrt[n]{x_1.x_2 \ldots .x_n}$.
- It is used to find the mean of the values which are exponential in nature, compound annual growth rates, interest rates, etc.

Harmonic Mean (HM)

- HM of *n* numbers is the reciprocal of the arithmetic mean of the reciprocals of all the numbers. Since reciprocals are used twice in the calculation, the effect of taking reciprocals is nullified.
- $HM = \dfrac{1}{\frac{1}{n}\Sigma\left(\frac{1}{x_i}\right)}$.
- It is used to find averages of different rates of the same event at different times or places.
- It minimizes the effect of extreme values in a set of data while drawing the averages.

Quartiles

- Quartiles are also positional averages when data are arranged in ascending or descending order. There are three quartiles—Q_1, Q_2 and Q_3.
- Q_1 includes the first 25%, Q_2 includes the first 50% and Q_3 covers the first 75% of the total arranged data.
- Q_2 and median is the same.
- Useful when merit score-based decisions to cover the first 25%, 50% or 75% are to be considered.

Deciles

- Deciles are also positional averages when data are arranged in ascending or descending order. There are nine deciles—D_1, D_2. . . , D_9.
- D_1 includes the first 10%, D_2 includes the first 20% . . . and D_9 includes the first 90% of data.
- D_5, Q_2 and median are the same.
- Useful when merit score-based decisions are to be taken covering the first 10%, 20%, etc.

Percentiles

- Percentiles are also positional averages when data are arranged in ascending or descending order. There are 99 percentiles—P_1, P_2 . . ., P_{99}.
- P_1 includes the first 1%, P_2 the first 2%, . . . , P_{99} the first 99% of data.
- Useful when merit score-based decision is to be taken covering the first 1%, 2%, etc.

It may be noted that P_{50}, D_5, Q_2 and median is the same and can be seen in Figure 13.1.

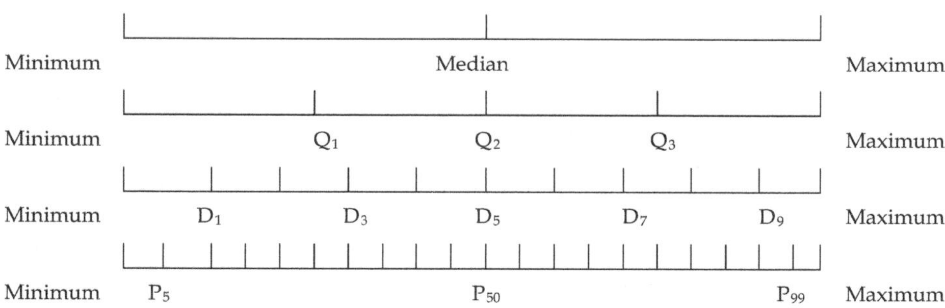

Figure 13.1 Depiction of Situation Where Median, Q_2, D_5 and P_{50} Coincide

13.3 Measures of Dispersion

Measures of dispersion are also known as measures of variability. Without the knowledge of variability in the dataset and using the measure of central tendency alone the spread of the dataset cannot be meaningfully determined. In fact, with the knowledge of the average along with knowledge of standard deviation, one can assess the nature of the dataset in terms of the central value and the spread of the data around the average. There are four basic measures of dispersion:

- Range
- Quartile Deviation
- Mean Deviation
- Standard Deviation and Variance

Range:

- Range is the difference between the highest value and lowest value in the dataset.
- It does not give the idea of the spread of data around a central value.
- Range of the dataset is used to get a proximate value of standard deviation where it appears in the formula to calculate sample size etc., in normally distributed data, range $\approx$ 6SD.

Quartile Deviation

- It helps to assess the spread of data about a measure of its central tendency. It gives the idea of a spread of 50% of data around the median.
- It is half of the difference between Q_3 and Q_1 or $QD = (Q_3 - Q_1)/2$.
- $Q_3 - Q_1$ is called inter-quartile range.

Mean Deviation

- It is the mean of absolute deviations of entire data with respect to the arithmetic mean.
- (Mean of difference of each data from AM, ignoring the sign).
- If $x_1, x_2, \ldots, x_n$ are *n* data, then $MD = \dfrac{\sum |X_i - \overline{X}|}{n}$.
- It does not ignore extreme values which may have importance in some cases and hence more useful in economics, commerce, business, etc.

Standard Deviation (SD)

- It is the widely used measure of dispersion or variability in a set of data.
- It gives a measure of the average of the deviations of the entire data in relation to the arithmetic mean. Low SD indicates that the values are close to the arithmetic mean and large SD indicates that the values are spread around the mean.
- SD of a population is denoted as σ and that of a sample as *S*, when calculated on the basis of 'n' or as s, when calculated on the basis of n−1.
- The variance is the square of SD and is calculated as variance $= \dfrac{\sum (X_i - \overline{X})^2}{n}$, then $\sqrt{\text{Variance}} = \text{SD}$.
- It has a wide range of applications in statistical inferences.

Coefficient of Variation (CV)

- It is the ratio of standard deviation (SD) to the arithmetic mean expressed in percentage. The higher the CV, the higher the variation in data and the lower the CV, the lower the variation in data around the mean.
- CV is calculated as $CV = \dfrac{\text{Standard Deviation}}{\text{Arithmetic Mean}} \times 100$.
- When two or more data series with different means and standard deviations are to be compared for within dataset consistency, then CV is the best measure for it. The dataset with the lowest CV is the most consistent.

Skewness

- It is the measure of the asymmetry of the probability distribution in relation to a symmetric normal distribution. For symmetric distribution, the left half of the histogram or frequency polygon will be a mirror image of its right half. For a skewed distribution the mean, median and mode will not be equal or the same.
- While mode is at the peak of the distribution, for skewed distribution the mean and mode can be on either side. For a positively skewed distribution, the mean and median will be on the right side of the mode, and for a negatively skewed distribution, those are on the left side of the mode.

- The skewness measures the departure from normality of the distribution of the dataset and is measured as skewness $= \dfrac{\sqrt{n}\sum_{i=1}^{n}\left(X_i - \bar{x}\right)^3}{\left(\sum_{i=1}^{n}\left(X_i - \bar{x}\right)^2\right)^{\frac{3}{2}}}$.
- The measurement of skewness can be positive, negative and zero. For positively skewed distribution, the longer tail is on the right, and for negatively skewed distribution, the longer tail is on the left side.

Kurtosis

- It measures the peakness or flatness of the distribution in relation to the bell-shaped normal curve.
- Kurtosis is a measure of flatness or peakness of the distribution in relation to its overall shape.
- The kurtosis of a standard normal distribution is three; it is called mesokurtic.
- Distribution with kurtosis, < 3 is said to be platykurtic and distribution with kurtosis >3 is said to be leptokurtic. Distribution with more kurtosis means more data at tails compared to normal distribution.
- Kurtosis $= \dfrac{\sum_{i=1}^{n}\left(X_i - \bar{X}\right)^4}{\left(\sum_{i=1}^{n}\left(X_i - \bar{X}\right)^2\right)^2} - 3.$

Box and Whisker Plots (box plot)

It is a visual device made using the quartiles of the dataset by following these steps:

- Arrange the data from the lowest to the highest on a horizontal axis.
- Draw a box above the horizontal axis in such a way that the left end of the box coincides with Q_1 (first quartile) and the right end with Q_3 (third quartile).
- The box is divided into two parts by a vertical line marking Q_2 (second quartile/median).
- Draw a horizontal line from the left end of the box to the point at the smallest value of the dataset and another line from the right end of the box to the point at the largest value of the dataset. These lines are called whisker lines. Figure 13.2 shows what the box and whisker plot looks like.

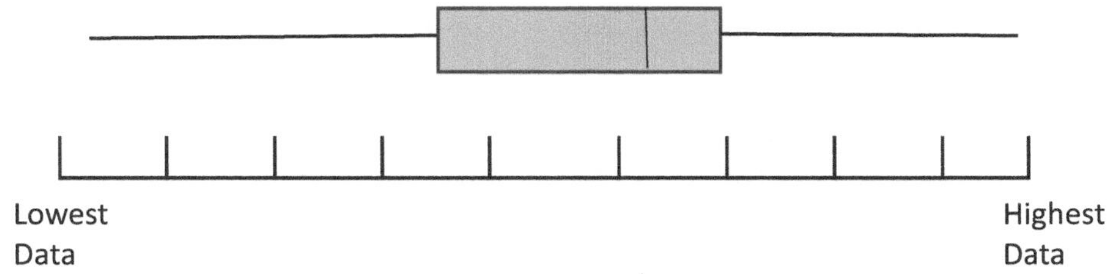

Figure 13.2 Box and Whisker Plot

- The box and whisker plot give an idea about the nature and spread of the data, focusing on the concentration of data and also the nature of symmetry of the dataset.
- Most of the statistical software draws the whisker plot to assess the nature of the dataset being analyzed and also locate the outliers, if any, in the dataset.

Outliers in the Dataset

- Outliers are usually very small and very large values in the dataset.
- Any data $X < Q_1 - 1.5$ times the interquartile range (IQR) or $X > Q_3 + 1.5$ times IQR in a set of data is considered as an outlier, where Q_1 and Q_3 are the first and third quartiles and $IQR = (Q_3 - Q_1)$.

13.4 Concept of Probability

Probability is associated with events that are the outcome of random experiments. Random experiment is an experiment that can be repeated under homogenous conditions, the results of which are one of the various possible outcomes/combinations of outcomes, but not exactly predictable. A random event is a set of favourable outcomes out of all possible outcomes. Every event has a probability. Example: the outcome of tossing a coin, sex of a newborn child, outcome of the throw of one dice/two dice, etc. There are three approaches to defining probability.

- Classical approach
- Empirical approach
- Axiomatic approach

13.4.1 Classical Probability

Let A random experiment have 'N' possible outcomes/cases out of which 'm' are favourable to the event 'A'. Then the probability of event A is denoted as P(A) and

$$P(A) = \frac{\text{Favorable No. of cases of A}}{\text{Total number of possible cases}} = \frac{m}{N} \quad 0 \le P(A) \le 1.$$

Example: If 85 out of 100 students pass an exam, the probability of passing $= \frac{85}{100} = 0.85$.

13.4.2 Empirical Probability

Empirical probability is the limiting value of classical probability, as the number of trials becomes large:

$$P(A) \Rightarrow \text{limit of } m/N \text{ as } N \to \text{infinity.}$$

Example: If four out of ten tosses of a coin are heads, the probability of getting heads = 0.4.
If 51 out of 100 tosses of a coin are heads, the probability of getting heads = 0.51.
If 499 out of 1000 tosses of a coin are heads, the probability of getting heads = 0.499.
If 5001 out of 10000 tosses of a coin are heads, the probability of getting heads = 0.5001.
Here, as the number of tosses increases, the probability of getting heads approaches 0.5.

13.4.3 Axiomatic Probability

It was introduced by A.N. Kolmogorov (1934). The probability of an event A in a sample space S of a random experiment satisfies the following axioms:

$$P(A) \geq 0$$

$$P(S) = 1$$

$P(A_1 + A_2 + \ldots + A_k) = P(A_1) + P(A_2) + \ldots + P(A_k)$, when $A_1, A_2, \ldots, A_k$ are disjoint events in S. Let S be a sample space having N number of possible outcomes, that is, n(S) = N. Let the number of sample points favourable to event A be n(A).

Then $P(A) = n(A)/n(S) = n(A)/N$.

13.5 Univariate Probability Distribution

There are many statistical probability distributions to calculate probabilities of specific events. These events can relate to discrete variables or continuous variables. All countable numbers like the number of family members, number of births, deaths, etc. fall under discrete variables. All measurements like height in cm, weight in kg, age in years, etc. are continuous variables. Some knowledge about the following two discrete and one continuous distribution is important in many statistical applications in research.

Binomial Distribution (Discrete Probability Distribution)
Poisson Distribution (Discrete Probability Distribution)
Normal Distribution (Continuous Probability Distribution)

13.6 Binomial Distribution

Let a random experiment is repeatedly performed a fixed number of times, say *n* times or *n* trials. The experiment has only two disjoint outcomes, say success and failure. All trials are independent; the preceding or succeeding outcome of the trial has nothing to do with the current outcome. Let the probability of success = p and that of failure is q = 1 − p, then the probability of *x* success in *n* trials is given by $p(x) = {}^{n}C_{x}p^{x}q^{n-x}$ where, x = 0, 1, 2, . . . , n. The mean of binomial distribution = *n*p and variance of binomial distribution = *n*pq and variance < mean.

While dealing with proportions having two disjoint cases like success or failure, pass or fail, male or female, etc. binomial distribution is the most appropriate one to calculate probabilities.

13.7 Poisson Distribution

It was derived in 1837 by Simeon D. Poisson. It is a limiting case of binomial distribution when *n* is very large $(n \to \infty)$, p probability of success is very small $(p \to 0)$ and *n*p = m is a definite number. Under these conditions, Binomial distribution tends to Poisson distribution. The probability function of Poisson distribution is given by:

$$P(X = x) = \frac{e^{-m}m^{x}}{x!} \quad x = 0,1,2,\ldots \text{ and } x! = x.(x-1)\cdot(x-2)\cdot\ldots:2.1$$

Mean = variance = m for the Poisson distribution.

Here np = m, hence, $p = \frac{m}{n}$ and $q = 1 - \frac{m}{n}$.

Number of success	probability
0	$e^{-m}.m^0/0! = e^{-m}$
1	$e^{-m}.m^1/1!$
2	$e^{-m}.m^2/2!$
3	$^{-m}.m^3/3!$

Example: Number of printing mistakes in a book, number of twins delivered in a hospital, etc.

13.8 Normal Distribution

- It is a continuous probability distribution.
- Derived by Karl Friedrich Gauss in 1733. The pdf of ND is given by:

Probability density function (pdf) $p(x) = \frac{1}{(\sqrt{2\pi}).\sigma} e^{-1/2\left(\frac{x-\mu}{\sigma}\right)^2} \quad -\infty \leq x \leq \infty.$

- π = 22/7 = 3.14 e = 2.718.
- Parameters of a normal distribution are μ = mean and σ = SD.
- ND has a bell-shaped curve and probability range with respect to mean ± SD (Figure 13.3).

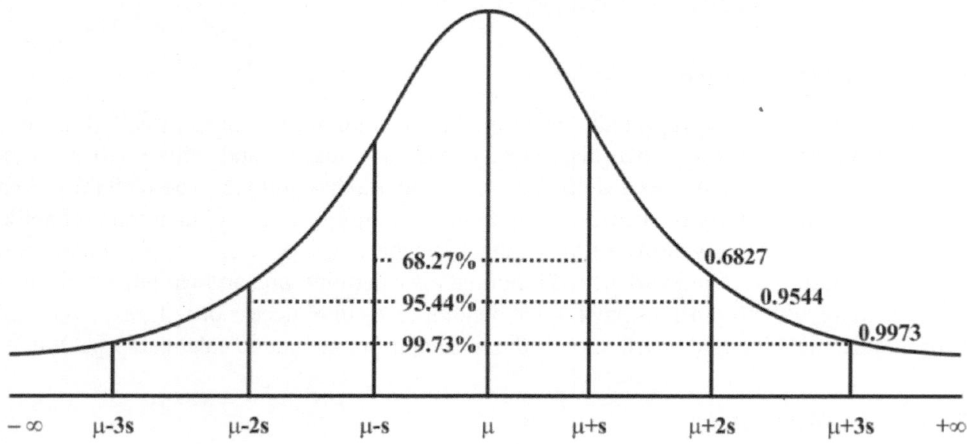

Figure 13.3 Normal Distribution Curve with Probability Range

Properties of Normal distribution:

- The graph of the curve is bell-shaped.
- The curve is symmetrical on the X-axis at mean (μ).
- For ND, mean = median = mode = μ.
- Area of curve before mean = area of curve after mean = 0.5.

- Coefficient of skewness (asymmetry) = 0.
- Coefficient of kurtosis (flatness) = 0.
- Theoretical range of X values is -∞ to +∞, practically range = 6σ.
- Maximum probability is at X = μ and $P = \frac{1}{\sqrt{2\pi}}.\sigma$.
- ND is unimodel.
- Pr (μ−σ ≤ X ≤ μ+σ) = 0.6827.
- Pr (μ−2σ ≤ X ≤ μ+2σ) = 0.9544.
- Pr (μ−3σ ≤ X ≤ μ+3σ) = 0.9973.

Standard Normal Distribution:

- If X follows N(μ,σ), then Z = (X − μ)/σ, follows N(0,1).
- $P(Z) = \frac{1}{\sqrt{2\pi}}.e^{-\frac{z^2}{2}}$.

  P(-1 ≤ z ≤ +1) =0.6826
  P(-2 ≤ z ≤ +2) = 0.9544
  P(-3 ≤ z ≤ +3) = 0.9973
  P(-1.96 ≤ z ≤ +1.96) = 0.95
  P(-2.58 ≤ z ≤ +2.58) = 0.99

  Figure 13.4 shows a standard normal distribution curve.

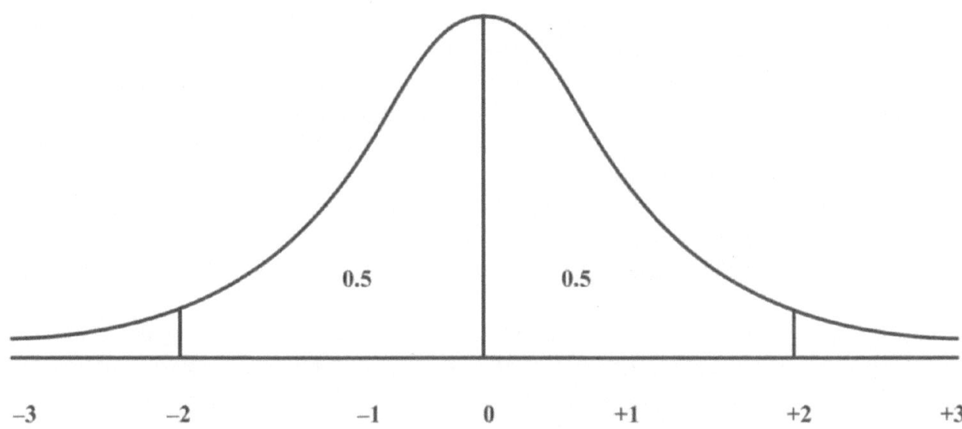

Figure 13.4 Standard Normal Distribution Curve with Probability Range

Some Important Points

- Measure of central tendency is a central representative value for the whole set of data that is either a calculated value or a positional average. Mean, median and mode are the most widely used measures of central tendency.
- AM of $x_1, x_2, \ldots, x_n$ is denoted as $\overline{x} = \frac{\text{Sum of all values}}{\text{n}} = \frac{\Sigma X_i}{n}$.

- The median is the middle observation for a dataset sorted in ascending or descending order. The median divides the data into a higher half and a lower half and hence is a positional average.
- Mode is the most repeated value in a set of data.
- GM of $x_1, x_2, \ldots, x_n$ is $GM = \sqrt[n]{x_1 . x_2 \ldots . x_n}$.
- HM of n numbers is the reciprocal of the arithmetic mean of the reciprocals of all the numbers.

$$HM = \frac{1}{\frac{1}{n}\Sigma\left(\frac{1}{x_i}\right)}$$

- Measures of dispersion are also known as measures of variability. It is the measure of the spread of the data around an average.
- Range is the difference between the highest value and lowest value in the dataset.
- Quartile deviation is half of the difference between Q_3 and Q_1 or $QD = (Q_3 - Q_1)/2$.
- Mean deviation is the mean of absolute deviations of the entire data with respect to arithmetic mean.
- Standard deviation gives a measure of the average of the deviations of the entire data in relation to the arithmetic mean. A low SD indicates that the values are close to the mean and a large SD indicates that the values are spread around the mean.
- SD of a population is denoted as σ and that of a sample as S or s.
- The variance is calculated as $\text{Variance} = \frac{\sum(X_i - \bar{X})^2}{n}$, then $\sqrt{\text{Variance}} = SD$.
- Quartiles are also positional averages when data is arranged in ascending or descending order. There are three quartiles—Q_1, Q_2 and Q_3. Q_1 includes the first 25%, Q_2 includes the first 50% and Q_3 covers 75% of the arranged data. Q_2 and median are the same.
- Deciles are also positional averages when data is arranged in ascending or descending order. There are nine deciles—$D_1, D_2, \ldots, D_9$. D_1 includes the first 10%, D_2 includes the first 20%, ... and D_9 includes the first 90% data.
- Percentiles are positional averages when data is arranged in ascending or descending order. There are 99 percentiles—$P_1, P_2, \ldots, P_{99}$.
- P_1 includes the first 1%, P_2 the first 2%. . . , P_{99} the first 99% of data. P_{50}, D_5, Q_2 and median are the same.
- Skewness measures the departure from normality of the distribution of the dataset.
- Kurtosis is a measure of the shape of the tail of the distribution in relation to its overall shape.
- An outlier in a dataset are very large or very small values in a set of data which can be ascertained using whisker plots based on first and third quartiles.
- Probability is associated with events which are the outcome of random experiments.
- A random event is a set of favourable outcomes out of all possible outcomes. Every event has probability.
- When A random experiment has 'N' possible outcomes/cases out of which 'm' are favourable to the event 'A'. Then the probability of event A is denoted as P(A) and $P(A) = \frac{\text{Favorable No. of cases of A}}{\text{Total number of possible cases}} = \frac{m}{N}$.
- **Binomial distribution:** If probability of success = p and that of failure is q = 1 − p, then the probability of x success in *n* trials is given by $p(x) = {}^nC_x p^x q^{n-x}$ where x = 0, 1, 2, . . . , n.

The mean of binomial distribution = *n*p and variance of binomial distribution = *n*pq and variance < mean.

- **Poisson distribution:** The probability function of Poisson distribution is given by $P(X=x)=\frac{e^{-m}m^{x}}{x!};x=0,1,2,\ldots$; mean = variance = m for the Poisson distribution.
- The probability density function of normal distribution is given by $f(x)=\frac{1}{\sqrt{2\pi\sigma}}e^{-\frac{1}{2}\left(\frac{X-\mu}{\sigma}\right)^{2}}$.
- The graph of the curve is bell-shaped; the curve is symmetrical at x = μ; for ND, mean = median = mode = μ; area of the curve before mean = area of the curve after mean.
- For ND, coefficient of skewness (asymmetry) = 0; coefficient of kurtosis (flatness) = 0; theoretical range of X values is -∞ to +∞, practically range = 6σ, maximum probability is at X = μ and is equal to $P=\frac{1}{\sqrt{2\pi}}.\sigma$ and ND is unimodel.
- $\Pr(\mu-\sigma\leq X\leq\mu+\sigma)=0.6827$; $\Pr(\mu-2\sigma\leq X\leq\mu+2\sigma)=0.9544$; $\Pr(\mu-3\sigma\leq X\leq\mu+3\sigma)=0.9973$.
- Put $Z=\frac{X-\mu}{\sigma}$, then the distribution of Z is standard normal distribution (SND) with N(0,1).
- $\Pr(-1\leq Z\leq+1)=0.6826$; $\Pr(-2\leq Z\leq+2)=0.9544$ and $\Pr(-3\leq Z\leq+3)=0.9973$.
- The standardized Z values are used to solve problems of normal distribution with given mean and SD.

Suggested Readings

Agresti, A., *Categorical data analysis*, 2nd ed., John Wiley & Sons, New York, 2002.

Bourke, G. J., J. McGilvary, and L. E. Daly, *Interpretation and uses of medical statistics*, Blackwell Scientific, London, 1985.

Daniel, W. W., *Biostatistics: basic concepts and methodology for the health sciences*, Wiley Publications, John Wiley & Sons. Reprint by Wiley India (P) Ltd, New Delhi, 2010.

Das, R., and P. N. Das, *Instant medical biostatistics*, Ane Books, New Delhi, 2009.

Dunn, O. J., *Basic statistics*, Wiley, New York, 1984.

Fisher, R. A., *Statistical methods for research workers*, 14th ed., Hafner Publishing Company, New York, 1973.

Gupta, S. C., *Fundamentals of statistics*, Himalaya Publishing House, Mumbai, 1981.

Gut, A., *An intermediate course in probability*, Springer-Verlag, New York, 1995.

Haberman, S., *The analysis of frequency data*, University of Chicago Press, Chicago, IL, 1974.

Hays, W. L., *Statistics for the social sciences*, 3rd ed., Holt, Rinehart, and Winston, New York, 1981.

Larsen, R. J., and M. L. Marx, *An introduction to mathematical statistics and its applications*, 2nd ed., Prentice-Hall, Englewood Cliffs, NJ, 1986.

Mahajan, B. K., and A. B. Khankal, *Methods in biostatistics for medical students and research workers*, 7th ed., Jaypee, New Delhi, 2010.

Muzumdar, R. D., A. P. Kulkarni, and J. P. Baride, *Manual biostatistics*, 1st ed., Jaypee Brothers, New Delhi, 2003.

Rees, D. G., *Essentials of statistics*, Chapman & Hall, London, 1989.

Rice John, A., *Mathematical statistics and data analysis*, 2nd ed., Duxbury, Belmont, CA, 1995.

Sharma, A. K., *The textbooks of elementary statistics*, Discovery Publishing House, Delhi, 2005.

Snedecor, G. W., and W. G. Cochran, *Statistical methods*, 7th ed., Iowa University Press, Ames, IA, 1980.

Woolson, R. F., *Statistical methods for the analysis of biomedical data*, Wiley, New York, 1987.

14 Correlation and Regression Analysis

14.1 Introduction

When we consider two or more variables at a time for their association or relationship, we use techniques like correlation and regression analysis. In many of the research methods and designs, we collect data on the stated dependent/outcome variable (Y) as well as on a number of independent/explanatory variables (Xs). Often researchers may be interested in examining the relationship of Y with X and also the extent of association among X explanatory variables.

14.2 Correlation Analysis

The correlation is a measure of the strength of association as well as the direction of association between variables. Numerically, the value of the linear correlation coefficient also known as Karl Pearson correlation coefficient lies between −1 and +1. The value of correlation coefficient −1 to zero implies a negative correlation or association and that between 0 to +1 implies a positive correlation. As the correlation is closer to one from either side indicates a strong association and closer to zero from either direction means a weak association. The correlation between dependent and independent variable and also between theoretically important independent variables can be meaningful quantitative technique to confirm the emerging results of the research. Correlation can be positive or negative as well as linear or non-linear. The magnitude of linear correlation coefficient and also the direction of correlation between variables can throw light on the nature of association between variables.

In most of the study areas, there are many variables having an association of one with the other. Correlation analysis of two or more variables helps to assess the strength as well as direction of association between the variables.

- Correlation can be linear or nonlinear (Y, Vs, X associated in linear or non-linear form).
- Correlation can be simple (r_{yx}) or multiple ($r_{Y.X1X2 \cdots Xk}$) or partial $r_{y.X1/X2 \cdots Xk}$).

Generally, the population correlation coefficient is denoted as 'p' and the sample correlation coefficient is denoted as 'r'.

14.2.1 Methods of Studying Correlation

- Scatter Diagram Method
- Karl Pearson Coefficient of Correlation
- Spearman's Rank Correlation
- Kendall's Tau Correlation

DOI: 10.4324/9781003527183-14

(i) **Graphical Method of Correlation analysis: Scattered diagram (Figure 14.1)**

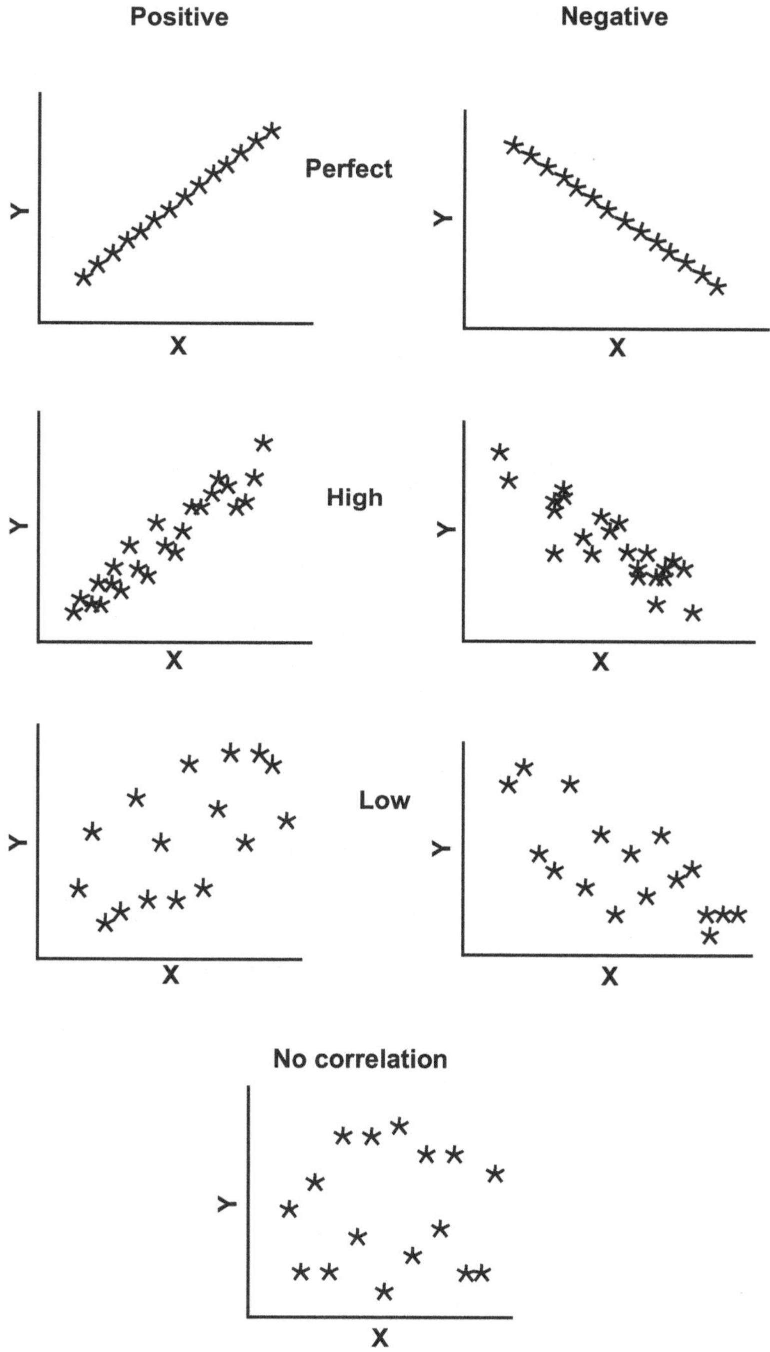

Figure 14.1 Scattered Diagram Method of Studying Correlation

(ii) **Karl Pearson Coefficient of Correlation**

For the calculation of Karl Pearson coefficient of correlation, the pairwise data of variable X and variable Y must be available. The Karl Pearson coefficient of correlation is calculated as:

$$\mathrm{r} = \frac{Cov.(x,y)}{\mathrm{SDx} \times \mathrm{SDy}} = \frac{\sum x_i y_i}{\sqrt{\sum x_i^2}\sqrt{\sum y_i^2}}$$

Where SDx is the standard deviation of X and SDy is the standard deviation of Y

$xi = Xi = \bar{X}$ and $yi = Yi - \bar{\mathrm{Y}}$.

(iii) **Rank Correlation (Spearman's Correlation Coefficient)**

When numerical data of X and Y are not available, instead the ranks of X and Y for the study units are available, then it is possible to calculate the rank correlation coefficient. It is calculated using the formula:

$$\mathrm{r} = 1 - \frac{6\sum d^2}{\mathrm{n}\left(n^2 - 1\right)}$$

Where, d = difference between ranks of attributes for each unit of study and n = number of pairs of observations.

(iv) **Rank Correlation (Kendall's Tau Correlation Coefficient)**

It is a nonparametric approach to measuring correlation coefficient using rank date. It varies from zero to one. Zero means no relation and one means perfectly related. For example, two teachers assigned ranks to ten students based on their knowledge. Each student having two ranks assigned by each teacher.

Arrange the students in the ascending order of rank assigned by the teacher one.

Mention the rank given by the second teacher against the name of each student.

Calculate the difference between the ranks for each student.

Count zero-difference students (A) and non-zero-difference students (B).

Kendall's Tau $\mathrm{r} = (\mathrm{A} - \mathrm{B}) / (\mathrm{A} + \mathrm{B})$.

The data must be in an equal number of pairs in order to calculate 'r'.

14.2.2 Properties of 'r'

- The value of r lies between −1 and +1.
- Correlation coefficient between two variables can be +ve or −ve.
- Example: Correlation coefficient between salt intake and BP is +ve.
- Correlation between BMI and physical work hours is −ve.

- If x and y are independent, then r = 0.
- If X and Y are perfectly correlated, then r = 1.
- There can be spurious or non-sense correlation when X and Y are influenced by a third variable, then r_{xy} (correlation coefficient) may be high, but it may be meaningless.

14.2.3 Partial Correlation

The partial correlation coefficient is the correlation between two variables keeping the effect of other variables constant. Let there be three variables X_1, X_2 and X_3. There can be three partial correlations:

- Partial correlation coefficient between X_1 and X_2, keeping X_3 constant ($r_{12.3}$).
- Partial correlation coefficient between X1 and X3, keeping X2 constant ($r_{13.2}$).
- Partial correlation coefficient between X2 and X3, keeping X1 constant ($r_{23.1}$).

Based on a simple correlation r_{12}, r_{13} and r_{23} the partial correlation coefficient can be calculated:

$$r_{12.3} = \frac{r_{12} - (r_{13})(r_{23})}{\sqrt{(1 - r_{13}^2)(1 - r_{23}^2)}}$$

$$r_{13.2} = \frac{r_{13} - (r_{12})(r_{23})}{\sqrt{(1 - r_{12}^2)(1 - r_{23}^2)}}$$

$$r_{23.1} = \frac{r_{23} - (r_{12})(r_{13})}{\sqrt{(1 - r_{12}^2)(1 - r_{13}^2)}}$$

Let X_1 be the number of tourists daily visiting a tourist centre for a fortnight, X_2 be the day temperature and X3 be the daily number of coffees sold at the centre, these three variables can have intercorrelations. The correlation $r_{12.3}$ is the partial correlation between the number of tourists visiting and daily temperature keeping the sale of coffee a constant $r_{13.2}$ is the partial correlation between the number of tourists visiting and the sale of coffee keeping the day temperature constant and $r_{23.1}$ is the partial correlation between day temperature and coffee sold keeping the number of tourists visiting a constant.

14.2.4 Multiple Correlation

The correlation of variable Y with joint effect of X_1, X_2, . . ., X_k is called multiple correlation and is denoted as $r_{Y.X1X2\ldots Xk}$. It is the square root of the coefficient of determination in multiple regression analysis (R^2).

14.2.5 Non-linear Correlation

The correlation of Y with X can be non-linear also and then it is called curve linear correlation, depending upon the nature of the curve formed by the plotted data on the X-Y plane.

14.3 Regression Analysis

The statistical method of establishing functional relationships between variables is called regression analysis.

$Y = f(x)$ is a simple regression model with one independent variable X and a dependent variable Y.

$Y = a + bX$ is a simple linear regression where 'a' is the intercept of the line and 'b' is the slope of the line; a and b are the two basic constants of the line. The corresponding regression model is $Y = a + bX + e$ where 'e' is the error term. The values of a and b are estimated using a given set of data of X and Y using ordinary least squares (OLS) or other methods of estimation. Various regression models are shown in Table 14.1.

Table 14.1 Regression Models and Their Mathematical Expressions

Regression	*Model*
Linear	$Y = a + bX$
Exponential	$Y = ab^X$
Logarithmic	$Y = a + b \log X$
Quadratic	$Y = aX^2 + bX + c$

The regression analysis means estimating functional relationships between the dependent variable and independent variables. In regression analysis the relationship of type Y = f(X) is established and used for forecasting and prediction purposes. The regression equation can be linear or non-linear, simple or multiple type.

The statistical method of establishing functional relationships between variables is called regression analysis.

Simple regression, Y = F(X) where, Y = dependent variable; and X= independent variable.
Multiple regression, $Y = F(X_1, X_2, \ldots X_k)$.

Table 14.2 shows various examples of dependent and independent variables in simple and multiple regressions. The first three are examples of simple regression and the last two are examples of multiple regression.

Table 14.2 Examples of Dependent and Independent Variables in Simple and Multiple Regressions

S. No.	*Y (Dependent)*	*X (Independent)*
1.	Blood pressure	Salt intake
2.	Blood sugar level	Sugar intake
3.	Number of children in HH	Literacy level of mother
4.	Body weight	Food intake and physical work hours
5.	Marks obtained	Per day hours of study and monthly attendance

14.3.1 Regression Types

When the value of Y increases or decreases at a constant rate with a change in the value of X, the regression of Y on X will be linear. When the value of Y increases at a higher rate with a change in the value of X, the regression of Y on X will be exponential. When the value of Y increases at a lower rate with a change in the value of X, the regression of Y on X will be log-linear. However, when the value of Y initially increases with the increase in the value of X, and after reaching a maximum the value of Y starts declining with increasing the value of X, the regression of Y on X will be quadratic, with the coefficient of X^2 negative. Similarly, when the value of Y initially decreases with the increase in the value of X, and after reaching a minimum the value of Y starts increasing with increasing the value of X, the regression of Y on X will be quadratic, with the coefficient of X^2 positive. These situations are shown in Figure 14.2.

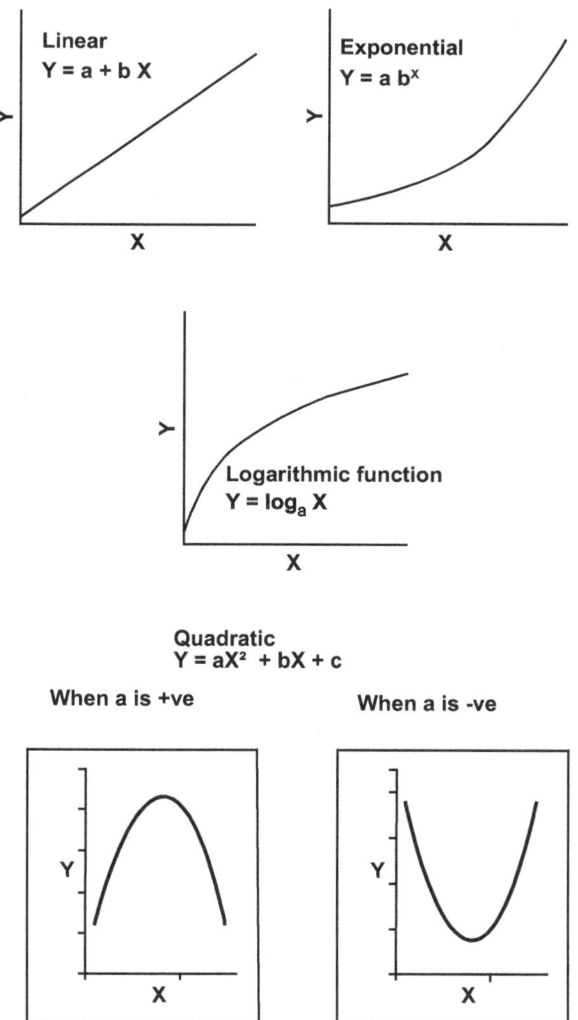

Figure 14.2 Types of Regression

Different situations of linear regression: A linear equation of the form $Y = a + bX$ has two constants, that is, a (intercept) and b (slope). Different situations with respect to the signs of 'a' and 'b' are given in Figure 14.3.

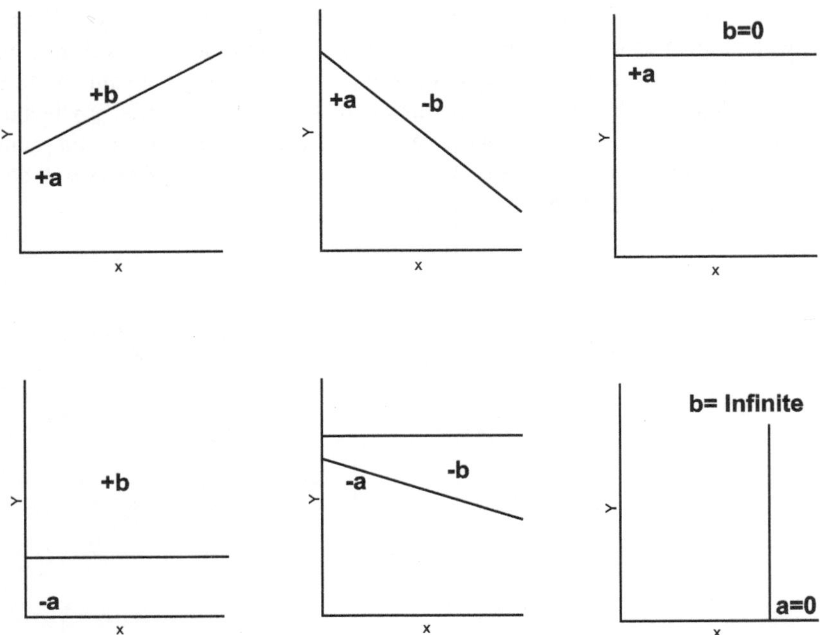

Figure 14.3 Forms of Line in Linear Regression

Forms of Line

Simple $y = f(x)$ $y = a + bx$ linear.
Multiple $y = f(x_1, x_2, \ldots x_k)$ $y = a + b_1x_1 + b_2x_2 + \ldots + b_kx_k$ linear.

14.3.2 Steps in Regression Analysis

(i) State the model.
(ii) Estimate parameters of the model.
(iii) Test the statistical significance of estimated parameters.
(iv) Assess the validity/strength of the model (using the coefficient of determination (R^2).
(v) Make statistical inferences.

14.3.3 Error Term in the Model

Let $Y = A + BX$, be the simple linear regression existing in the population as a line representing the scatter of (X, Y) points on the plain. In other words, corresponding to each point X_i on the X-axis, there is a Y_i, on the (XY) plane and also there is a point $\widehat{Yi}$ on the regression line. The difference between Y_i and $\widehat{Yi}$ is called the error and is denoted as 'u_i' for the population and e_i for the corresponding sample as shown in Figure 14.4

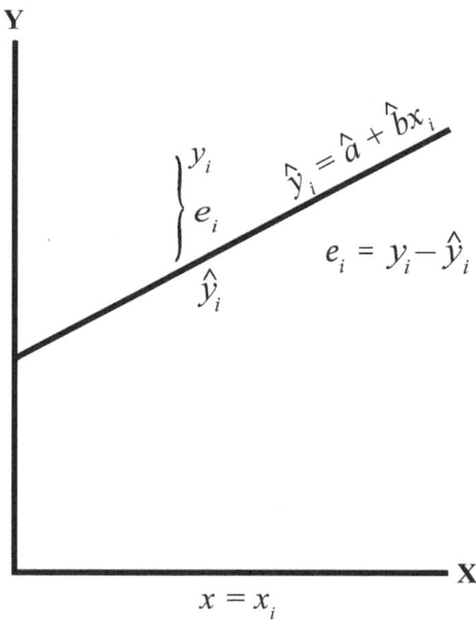

Figure 14.4 Error Term in Regression

Hence the actual relationship in population is $Y_i = A + BX_i + u_i$ and that in the sample is $y_i = a + bx_i + e_i$, where 'a' and 'b' are estimates of the population regression coefficients A and B.

14.3.4 Ordinary Least Squares (OLS) Method

There are different methods for estimating the regression coefficients. The most widely used method is the OLS method. Assuming that $e_i = Y_i - \widehat{Yi}$, the OLS method is where A and B are estimated as 'a' and 'b' by minimizing Σe_i^2.

The OLS estimates of A and B have the desired properties of BLU (best, linear and unbiased) and hence are preferred over other methods of estimation.

Assumptions about error term in regression analysis using OLS method.

- U_i or e_i is a random variable which can take +ve, −ve or 0 values.
- The mean of the error term is '0'. That is, $E(u_i) = 0$.
- The error term has constant variance. That is, $Var(u_i) = E[u_i - E(u_i)]^2 = \sigma_{u^2}$. This assumption of the error term is known as the assumption of homoscedasticity and its violation is called the problem of **heteroscedasticity**.
- Error term follows a normal distribution with zero mean and constant variance, or u_i follows $N(0, \sigma_u^2)$.
- The error term for different values of X (u_i for X_i and u_j for X_j) is independent. In other words, $Cov(u_i, u_j) = 0$. Its violation is called the problem of **autocorrelation.**

Other assumptions for the application of the OLS method

- In the case of multiple regression the explanatory variables $X_1, X_2, \ldots, X_k$ are not highly correlated to each other. Its violation is called the problem of **multi-collinearity.**
- The covariance between explanatory variable X_j and error term u_i are independent. In other words, $Cov(u_i, X_j) = 0$ for i = 1, 2, . . ., n and j = 1, 2, . . ., k.
- The mathematical relationship being estimated is correctly identified.
- The relationship is correctly specified without any specification error.
- The observations are correctly measured.
- There is no aggregation error in the case of variables obtained as the aggregate of measurements.

14.3.5 Violation of Assumptions

The problems of heteroscedasticity, auto-correlation and multi-collinearity have consequences in estimating the regression parameters and their standard errors (SE) for testing the statistical significance.

(i) **Problem of Heteroscedasticity:**

The violation of the assumption that the error term has constant variance, (i.e. $Var(u_i) = E[u_i - E(u_i)]^2 = \sigma_u^2$) leads to the problem of heteroscedasticity in regression analysis.

Consequences of heteroscedasticity:

- The SE formula for a test of significance of regression parameters is not applicable.
- The OLS estimate will not have minimum variance.
- The estimated regression coefficient can be unbiased, even when there is heteroscedasticity.
- The prediction of Y for a given X based on regression estimates, may have higher variance or prediction will be inefficient.

The Glejser test: Based on regression of |e| with explanatory variable in different power form. There are different tests for heteroscedasticity and solution to overcome the problem in a given set of data.

(ii) **Problem of Auto Correlation/Serial Correlation:**

It is the violation of assumption that $Cov(u_i, u_j) = 0$. In other words, the error term at a particular period is correlated with its own preceding values. Here $u_t = \rho u_{t-1} + v_t$ is known as first-order autocorrelation. A measure of first-order autocorrelation is assessed using autocorrelation coefficient which is calculated as:

$$r_{(e_t, e_{t-1})} = \frac{\sum_{t=1}^{n} e_t e_{t-1}}{\sqrt{\sum_{t=1}^{n} e_t} \times \sqrt{\sum_{t=1}^{n} e_{t-1}}}.$$

Reasons for autocorrelation:

- Omitted explanatory variable in the model.
- Misspecification of the form of relationship.

- Use of interpolated data in the model.
- Miss specification of the true random error.

Consequences of autocorrelation:

- Even when error terms are autocorrelated the estimated parameters of the model can be unbiased.
- With autocorrelated error terms, the OLS estimate of the variance of the parameter is likely to be more.
- The variance of the error term may be under-estimated.
- Prediction based on such estimates may be inefficient.

Test for autocorrelation:

The Von Neumann ratio test and the Durbin-Watson test can be used for the test of autocorrelation.

(iii) **Problem of Multi-Collinearity:**

It is a problem in multiple regression, that is, $Y = f(X_1, X_2, \ldots, X_k)$. Multi-collinearity is due to the violation of the assumption that explanatory variables X_i and X_j are not perfectly correlated ($r_{(X_i,X_j)} \neq 1$).

Reasons for multi-collinearity:

- Tendency of explanatory variables to move together over time.
- Due to the use of lagged variables in the model.

Consequence of multi-collinearity:

- When ($r_{(X_i,X_j)} = 1$ then the parameters of the model become indeterminate ($\frac{0}{0}$).
- On the contrary if $r_{(X_i,X_j)} = 0$, then there is no need to perform multiple regression analysis.

Detection:

Klein criteria: If $r^2_{(X_i,X_j)} \geq R^2_{Y.X_1,X_2,\ldots,X_k}$ then collinearity is harmful for the coefficient for X_i and X_j.

Remedial measures for multi-collinearity:

- Ignore the problem if it does not seriously affect the estimates.
- Drop unimportant variable out of two multi-collinear variables.
- Increase sample size in the study.
- Application of principle component analysis.

14.3.6 Model and Estimation of Simple Linear Regression

The linear equation is of the form, $y = a + bx$, a sale is a function of advertisement cost. The corresponding model is $Y = a + bx + e$, where e is the error term. The error is the difference between observed value of Y and estimated value of Y denoted as $\hat{Y}$. Corresponding to each value of Y there will be a $\hat{Y}$.

That is, $\hat{Y} = \hat{a} + \hat{b}$ Xi where $\hat{a}$ and $\hat{b}$ are the estimated values of a and b respectively. We can define ei as:

$$ei = Yi - \hat{Y}i.$$

14.3.6.1 Statistical Technique to Estimate 'a' and 'b'

The values for 'a' and 'b' are estimated using the principle of ordinary least squares using the above relationship.

$$ei = Yi - \hat{Y}i \text{ or } ei^2 = (Yi - \widehat{Yi})^2 \text{ or } ei^2 = [Yi - (\hat{a} + \hat{b}Xi)]^2.$$

By applying mathematical conditions for minimum value of ei^2, two equations in a and b are derived and by solving these equations the values of a and b are worked out as:

$$b = \frac{\sum xy}{\sum x^2} \text{ where}$$

$$a = \overline{Y} - b\overline{X}$$

$$x = X - \overline{X}$$

$$y = Y - \overline{Y}\ .$$

14.3.6.2 Validity of Model

The validity of the model depends on coefficient of determination (R^2) where

$$R^2 = \frac{\text{Explained variation in Y}}{\textit{Totalvariationin Y}} = \frac{\sum xy}{\sum y2},$$ which is the explanatory power of the model to

explain variation in Y due to that of X and $0 \leq R^2 \leq 1$.

Uses of Regression Model

- To assess the status of the situation.
- To forecast future values.
- To make policy decisions.

14.3.7 Multiple Linear Regression Analysis

For $Y = a + b_1X_1 + b_2X_2 + \ldots + bk_kX_k$ the coefficients are to be estimated for forecasting Y for given X values. Here it is assumed that:

- Dependent variable Y and independent variable X are metric variables.
- Regression Coefficients (a, b_1 b_2, . . ., b_k) are estimated using ordinary least squares (OLS) method.
- Regression coefficients (b_i for X_i) are weight/marginal value/rate of change of Y due to X_i.
- Assumptions of normality, linearity and equal variance followed.
- There are more than two explanatory variables associated with a dependent variable.

For example, the quantity demanded of a commodity is a function of its price (−ve association) and the income of the consumer (+ve association).

The body mass index (BMI) of a group of people is a function of calorie intake (+ve association) and physical work hours (−ve association).

Multiple Regression Types:

$Y = a + b_1x_1 + b_2x_2 + \ldots + b_kx_k$—multiple linear equation.

$Y = ax_1^{b1}x_2^{b2}\ldots x_k^{bk}$—multiple non-linear equation.

$Y = a + b_1x_1 + b_2x_2 + \ldots + b_kx_k + ei$—multiple linear regression model.

$Y = a_1^{b1}x_2^{b2}\ldots x_k^{bk}ei$—multiple non-linear regression model.

a, b_1, b_2, . . . b_k are estimated by minimizing $\sum ei^2$ w.r.t a, b_1, b_2, . . . b_k.

Two Explanatory Variable Model

$$Y_i = a + b_1X_{1i} + b_2X_{2i} + ei$$

$$ei = Y_i - a - b_1X_{1i} = b_2X_{2i}$$

$$\Sigma ei^2 = \Sigma\left(Y_i - a - b_1X_{1i} - b_2X_{2i}\right)^2$$

a, b_1, b_2 are estimated such that $\sum ei^2$ is minimum.

$$b_1 = \frac{\left(x_2^2x_1y - x_1x_2x_2y\right)}{x_1^2x_2^2 - \left(x_1x_2\right)^2}$$

$$b_2 = \frac{\left(x_1^2x_2y - x_1x_2x_1y\right)}{x_1^2x_2^2 - \left(x_1x_2\right)^2}$$

$$a = \bar{Y} - b_1\overline{X_1} - b_2\overline{X_2}$$

$$R^2 = \frac{b_1(x_1y) + b_2(x_1y)}{y^2}$$

Test of Significance

The estimated coefficients a, b_1 and b_2 are to be tested for their significance through a t-test.

$$t_a = \frac{\hat{a}}{SE\ of\ \hat{a}}$$

$$t_{b1} = \frac{\widehat{b1}}{SE\ of\ \widehat{b1}}$$

$$t_{b2} = \frac{\widehat{b2}}{SE\ of\ \widehat{b2}}$$

where $\hat{a}, \widehat{b1}$ and $\widehat{b2}$ are estimated values of a, b1 and b2.

Steps in Multiple Regressions

- State the model.
- Estimate a, b_1, b_2, etc.
- Test significance of a, b_1, b_2, etc.
- Calculate R^2 value (coefficient of determination).
- Test for significance of R^2.

14.3.8 ANOVA in Regression

ANOVA is applied in regression for testing the overall significance of regression (R^2). It is also useful to assess the improvement in fit due to additional explanatory variables.

Equality of regression coefficients from different samples can also assessed from ANOVA.

Format of ANOVA is given in Table 14.3.

Table 14.3 ANOVA for Regression Analysis

Source of variation	*SS*	*DF*	*MS*	*F*
Regression	RSS	$DF_r = k-1$	$MS_r = RSS/DF_r$	MS_r/MS_e
Error	ESS	$DF_e = n-k$	$MS_e = ESS/DF_e$	
Total	TSS	n−1		

14.3.9 Logistic Regression

Regression analysis is a complex and powerful statistical technique for the prediction of a dependent variable for any independent/explanatory variable based on an estimated relationship using known values of dependent and independent/explanatory variables. In normal regression analysis Y = F(X), the dependent variable is quantitative type (discrete or continuous), and the independent variables can be discrete, continuous, ordinal or dichotomous/nominal. Regression analysis enables us to predict the dependent variable and to know the rate of change of the dependent variable due to a unit change in the independent/explanatory variable or the elasticity coefficient of Y with respect to X. When we have to use qualitative/categorical variable as an independent variable in regression analysis, we use the technique of dummy variable (x = 0 or 1). Stepwise regression is one of the effective ways to select the most important independent variables out of a set of available independent variables. The forward or backward selection of independent variables is also possible in such situations.

Logistic regression is a special type of regression used when the dependent variable (Y) is non-metric (binary/dichotomous) in nature and independent variables are discrete, dichotomous or continuous in nature. Here, Y = 1 or 0 for the regression Y = F(X). The outcome variable can be cured or not cured, smoker or non-smoker, male or female, etc. Logistic regression is a technique in which the dependent/outcome variable is dichotomous.

Let us consider the simple linear regression model of the following type:

$$Y = a + bX + e, \tag{1}$$

Where a is the intercept, b is the slope and e is the error term. The parameters 'a' and 'b' are estimated using an ordinary least squares technique.

We know that at mean of X and Y, e = 0.

That is,

$$\mu_{y/x} = E(Y/X) = a + bX \tag{2}$$

When Y is dichotomous, it may take values of one or zero and the mean of Y will lie between zero and one.

Let p = Pr(Y = 1) or the probability of success, the ratio of, $\frac{p}{1-p}$ can take values between 0 and ∞ and

$\text{Ln}\left(\frac{p}{1-p}\right)$ can take values between $-\infty$ and $+\infty$.

$$\text{Hence } \text{Ln}\left(\frac{p}{1-p}\right) = a + bX. \tag{3}$$

Equation (3) above is called the logistic regression model and the transformation of p to $\text{Ln}\left(\frac{p}{1-p}\right)$ is called logit transformation. Equation (3) can also be written as

$$p = \frac{\exp(a + bX)}{1 + \exp(a + bX)} \tag{4}$$

where exp is the inverse or antilog of the natural logarithm.

The logistic regression model is widely used in health science research. Epidemiologists may use a logistic regression model to assess the probability or risk that a person will acquire a disease during some specified time interval during which he or she is exposed to risk factors related to that disease.

14.3.9.1 Case 1: Independent Variable Is Dichotomous

The simplest case for the application of logistic regression is when both dependent and independent variables are dichotomous. The values of the dependent/outcome variable (Y) indicate whether or not the person acquired a disease. The values of the independent variable (X) indicate the presence or absence of risk factors in the person. Hence both dependent and independent variables take the values of one or zero based on yes or no situations.

The data in such a situation can be presented in a 2 × 2 contingency table as given in Table 14.4

The analysis of these data can be done through odds ratio. According to the theory of probability, the odds for success is the ratio of the probability of success to the probability of failure. The odds ratio is a measure of how much greater or less the odds are for subjects possessing the

Table 14.4 2 × 2 Contingency Table

Dependent Variable	*Independent Variable*	
	Yes (1)	No (0)
Yes (1)	n_{11}	n_{10}
No (0)	n_{01}	n_{00}

risk factor to experience a particular outcome. The value of the odds ratio reflects the number of times the exposure to the risk factor can cause the case/disease as compared to those not exposed to the risk factor.

The computer package performing logistic regression generally provides the following:

- The intercept(a) of the linear relationship $Y = a + bX$.
- The slope (b) of the linear relationship $Y = a + bX$.
- The numerical value of the odds ratio.

Here the odds ratio = exp(b) which gives the number of times the exposure to the risk factor can cause the case or disease as compared to those not exposed to the risk factor.

14.3.9.2 Case 2: Independent Variable is Continuous

Here the dependent variable is dichotomous and the independent variable is continuous. Assuming that the logistic regression is worked out using computer software, the computer output will be similar to that shown previously.

The estimated equation will be $\widehat{yi} = \hat{a} + \hat{b} \cdot \mathrm{xi}$, where $\widehat{yi} = Ln\left[\dfrac{\hat{p}}{1-\hat{p}}\right]$ and $\hat{p}$ is the predicted probability of having the risk for the value of continuous variable xi.

14.3.9.3 Case 3: Multiple Logistic Regressions

Normally the researchers may be interested in a relationship of several independent variables with respect to a dependent variable. Here the dependent variable is dichotomous and independent variables are either continuous or a combination of discrete and continuous.

Multiple logistic regression model is of the following type:

$$\mathrm{Ln}\left(\frac{p}{1-p}\right) = \mathrm{a} + \mathrm{b}_1\mathrm{X}_1 + \mathrm{b}_2\mathrm{X}_2 + \ldots\ldots + \mathrm{b}_k\mathrm{X}_k.$$

Using the logit transformation we have

$$\mathrm{P} = \frac{\exp\left(a + b_1X_1 + b_2X_2 + \ldots + b_kX_k\right)}{1 + \exp\left(a + b_1X_1 + b_2X_2 + \ldots + b_kX_k\right)}.$$

14.3.9.4 Case 4: Polytomous Logistic Regression

The previous cases 1, 2 and 3 were for dichotomous dependent or outcome variables. There may be cases where the response/outcome variable may have a situation with multiple categories. For example, we may have outcomes as positive, negative and undetermined. In cases like the study of body mass index (BMI) the outcome may be underweight, ideal weight, overweight or obese. Such a situation is called an ordinal polytomous outcome or response. It is possible to have computer software-based logistic regression analysis for such cases also, though it is slightly complex.

Some Important Points

- Correlation analysis of two or more variables is to assess the strength as well as direction of association between the variables.
- Methods of studying correlation include the scatter diagram method, Karl Pearson coefficient of correlation and rank correlation.
- Scatter diagram is the graphical method of correlation analysis.
- Karl Pearson coefficient of correlation is calculated as $r = \frac{Cov(X,Y)}{SD_X \times SD_Y}$.
- The data of X and Y must be in an equal number of pairs. The value of r lies between −1 and +1.
- Correlation coefficient between two variables can be +ve or −ve. If x and y are independent, then r = 0. If X and Y are perfectly correlated, then r = 1. The correlation can be perfect, high or low depending upon the value of r. The correlation can be positive or negative depending upon the sign of r.
- The partial correlation coefficient is the correlation between two variables keeping the effect of other variables constant.
- The statistical method of establishing functional relationships between variables is called regression analysis. ·$Y = f(x)$ is a simple regression model with one independent variable X and a dependent variable Y. Multiple Regression model is of the type, $Y = F(X_1, X_2, \ldots X_k)$ with k dependent variables.
- Regression models can be linear or non-linear type, and the regression models also take forms accordingly.
- Error term in the regression model stands for the difference between the observed value and the corresponding estimated value on the regression line/curve $e_i = Y_i - \widehat{Yi}$.
- The error term is assumed to follow a normal distribution with '0' mean and constant variance. The error term at different values of explanatory variables is assumed to be independent.
- The explanatory variables (X_1, X_2, . . ., X_k) are assumed not to have high correlation for OLS application.
- Ordinary least squares (OLS) method is that method wherein the parameters of the regression model are estimated by minimizing Σei^2.
- The OLS estimates have the desired properties of BLU (best, linear and unbiased) and hence preferred over other methods of estimation.
- The violation of assumptions about error term and explanatory variables in the OLS method leads to consequences like heteroscedasticity (violation of constant variance), autocorrelation (violation of independent error term) and multi-collinearity (violation of not having a high linear correlation between explanatory variables) which affect the test of significance of estimated parameters.
- The validity of the model depends on the coefficient of determination (R^2) where $R^2 = \frac{\text{Explained variation in Y}}{\text{Total variation in Y}}$, which is the power of the model to explain variation in Y due to that of X and $0 \le R^2 \le 1$.

- Test of Significance of the estimated coefficients of the regression model is done using a t-test.
- ANOVA is applied in regression for testing (i) overall significance of regression (R^2), (ii) improvement in fit due to additional explanatory variable, (iii) equality of regression coefficients from different samples.
- Logistic regression is a special type of regression used when the dependent variable (Y) is non-metric (binary/dichotomous) in nature and independent variables are discrete, dichotomous or continuous in nature.

Suggested Readings

Cameron, A. C., and P. K. Trivedi, *Regression analysis of count data*, Cambridge University Press, Cambridge, 1998.

Daniel, W. W., *Biostatistics: basic concepts and methodology for the health sciences*, Wiley Publications, John Wiley & Sons, Reprint by Wiley India (P) Ltd., New Delhi, 2010.

Dielman, T. E., *Applied regression analysis for business and economics*, 2nd ed., Duxbury, Belmont, CA, 1996.

Dobson, A. J., *An introduction to generalized linear models*, 2nd ed., Chapman & Hall/CRC, Boca Raton, FL, 2002.

Draper, N. R., and H. Smith, *Applied regression analysis*, 2nd ed., John Wiley & Sons, New York, 1981.

Hocking, R. R., *Methods and applications of linear models: regression and the analysis of variance*, Wiley, New York, 1996.

Hosmer, D. W., and S. Lemeshow, *Applied logistic regression*, 2nd ed., John Wiley & Sons, New York, 2000.

Johnstone, I. M., and P. F. Velleman, 'The resistant line and related regression methods', *Journal of the American Statistical Association* 80: 1041–1054, 1985.

Koutsoyiannis, A., *Theory of econometrics*, Macmillan, London, 1976.

Mendenhall, W., and T. Sincich, *A second course in statistics: regression analysis*, 5th ed., Prentice Hall, Upper Saddle River, NJ, 1996.

Nagelkerke, N. J. D., 'A note on the general definition of the coefficient of determination', *Biometrika* 78(3): 691–692, 1991.

Neter, J., M. H. Kutner, C. J. Nachtsheim, and W. Wasserman, *Applied linear regression models*, 3rd ed., Irwin, Chicago, IL, 1996.

15 Statistical Inference—Parametric and Non-Parametric Tests

15.1 Statistical Inference

The process of making a decision about population parameters based on evidence given by a sample is called statistical inference. It has two parts. First, is the estimation of population parameters based on sample data and second is the testing of the significance of sample estimates for population parameters. The common assumptions for estimation and testing include assumptions about normality of population distributions, equality of variances and independence of samples. There are two types of testing—parametric and non-parametric. The definitions of important terms in parametric tests are:

Population: Aggregate of all units for a study having the focused features of the study is called a population.

Sample: A representative part of the population, normally selected randomly from the population.

Parameter: The statistical constants of the population is called parameters. For example, population mean (μ), population variance (σ^2) or standard deviation (σ), population correlation-coefficient (p), etc.

Statistic: The statistical constants of the sample like sample mean ($\bar{X}$), variance (S^2 or s^2) and correlation coefficient (r) are called statistics. These are the estimates for population parameters.

Test of Significance/Testing of Hypothesis: The inductive inference of deciding about population parameters based on sample statistics involves an element of risk—the risk in making wrong decisions. Minimization of such risk using the theory of probability is possible. Making decisions about population parameters using sample statistics by minimizing the risk is called the test of significance/testing of hypothesis. The theory of the test of significance was initiated by J. Neyman and E.S. Pearson.

Research Hypothesis: A research hypothesis is a statement of expectation or prediction of outcome or results through the conduct of research. It is in the form of a research question and its expected results. Normally, the research hypothesis resembles the alternative statistical hypothesis.

Statistical Hypothesis: A statistical hypothesis is a statement about population parameters (may or may not be true) which is tested based on evidence from a sample using the theory of probability. Normally it relates to the magnitude or relationship of population parameters.

Null Hypothesis (H_0): Null hypothesis is a hypothesis which is tested under the assumption that it is true. It is generally the hypothesis of no difference. It is drawn based on a neutral or null attitude by the researcher or decision maker.

Alternative Hypothesis (H_1): It is that hypothesis which is accepted if the null hypothesis is rejected.

DOI: 10.4324/9781003527183-15

Simple Hypothesis: If the hypothesis completely specifies the population, it is simple.

$$H_0 : \mu = \mu_0$$

Composite Hypothesis: If the hypothesis does not completely specify the population, then it is a composite hypothesis.

$$H_1 : \mu < \mu_0 ; \mu > \mu_0 ; \mu \neq \mu_0$$

Types of Errors: When we make inferences about population parameters based on evidence given by the sample, two types of errors are possible (Table 15.1).

First, the sample evidence may lead to rejection of a null hypothesis, when it is actually true. It is called **Type-I error.**

Second, the sample evidence may lead to acceptance of a null hypothesis, when it is actually false. It is called **Type-II error.**

Different types of error in statistical inference are given in Table 15.1.

Table 15.1 Types of Error in Statistical Inference

Situation of H_0 in population	*Decision on H_0 based on sample evidence*	
	Reject H_0	*Accept H_0*
H_0 True	Wrong (Type-I error)	Correct
H_0 False	Correct	Wrong (Type-II error)

Test Criteria: In a real-world situation, a Type-II error is more serious than a Type-I error. Both the errors cannot be reduced simultaneously. The test criteria fix the size of Type-I error at a low value and devise test criteria which minimize Type-II error.

α = probability of rejecting H_0/H_0 is true or probability of Type-I error = α (producer's risk).
β = probability of accepting H_0/H_0 is false or probability of Type-II error = β (consumer's risk).
Probability of accepting H_0 when H_0 is wrong + probability of accepting H_0 when H_0 is true = 1.
Probability of accepting H_0 when H_0 is true = 1. Probability of accepting H_0 when H_0 is wrong = $1-\beta$.

Power of Test and Level of Significance:

Power of test = $1-\beta$ = 1—Probability (Type-II error).

Level of Significance: The size of Type-I error, that is, probability rejecting H_0/H_0 is true = α. It is the probability of rejecting a true null hypothesis. It is generally kept at 5% ($p = .05$) or 1% ($p = .01$).

Confidence coefficient (%) = $(1-\alpha)$%.

Critical Region: It is the region of rejection in the probability space.

Critical Value or Significant Value: The value of the test statistic which separates the critical region and acceptance region is called critical value or significant value. It depends upon:

(i) The level of significance.
(ii) The nature of the alternative hypothesis (one-tailed or two-tailed).
(iii) Probability of Type-I error = α (producer's risk).
(iv) Probability of Type-II error = β (consumer's risk).

In real-world situations, Type-II error is more serious than Type-I error. Both the errors cannot be reduced simultaneously.

One-sided and Two-sided Tests: Depending upon the nature of the alternative hypothesis, a test can be one-sided or two-sided. If the alternative hypothesis is 'not equal to', then the alternative hypothesis can fall on either side of the curve and if it is 'greater than or less than' it falls either on the right side or the left side of the curve as shown (Figure 15.1 to Figure 15.3).

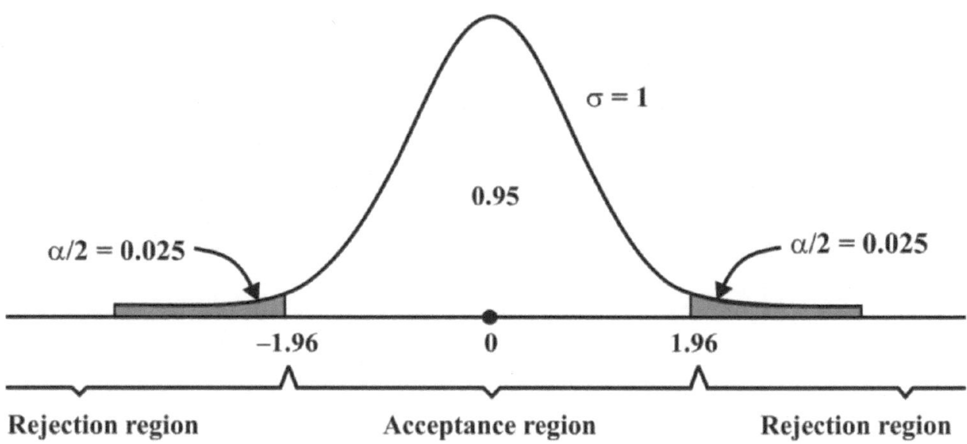

Figure 15.1 Acceptance and Rejection Region for Two-Sided Test

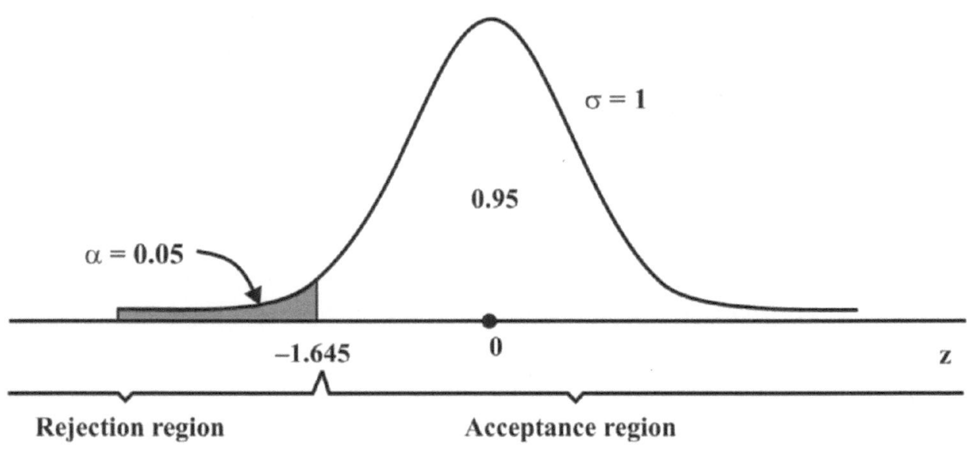

Figure 15.2 Acceptance and Rejection Region for One-Sided Left Test

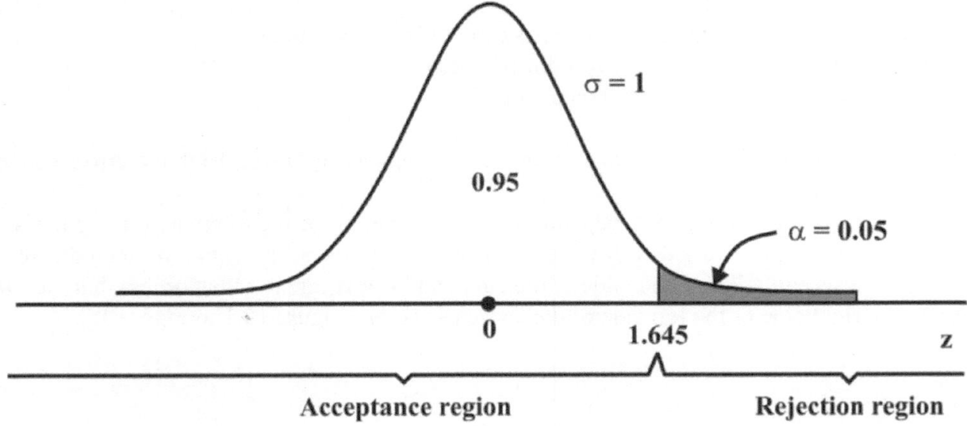

Figure 15.3 Acceptance and Rejection Region for One-Sided Right Test

15.2 Sampling Distribution

Let the size of the population be N and that of the sample be n. Then there can be $^{N}C_{n} = k$ samples. Let $t_1, t_2, \dots t_k$ be the values of the sample statistic 't'. The set of values of 't' is called the sampling distribution of statistic 't'. Like any other variable 't 'can also have mean and variance.

$$\text{Mean}(t) = E(t) = \bar{t}$$

$$\text{Var}(t) = 1/k \cdot \Sigma(t - \bar{t})^2$$

15.2.1 Standard Error (S.E.)

The standard deviation of the sampling distribution of a statistic is called standard error (S.E.)

$$\text{S.E.}(t) = \sqrt{\left[1/k\left(\Sigma(t - \bar{t})^2\right)\right]}.$$

Precision: The precision of a sample estimate for a population parameter is the reciprocal of the S.E. of the estimate.

$$\text{Precision of estimate } t = \frac{1}{S.E.(t)}.$$

15.2.2 Sampling Distribution of Mean

Let $X_1, X_2, \dots X_n$ be a random sample of size n from a normal population N (μ, σ), then the sample mean $\bar{X}$ is taken as an estimate of population mean μ, then,

$$\text{Variance }(\bar{X}) = \sigma^2 / n \text{ and S.E. }(\bar{X}) = \sigma / \sqrt{n}.$$

According to the central limit theorem, $Z = \dfrac{t - E(t)}{S.E.(t)}$ will follow N (0, 1).

15.2.3 Sampling Distribution of Proportion

Let A and B be two qualitative attributes and the entire population units (N) fall in either of these two attributes. Let there be 'a' units in A and 'b' units in B. That is, $a + b = N$ or $b = N - a$.

The proportion 'P_a' possessing attribute $A = a / N$.

The proportion P_b possessing attribute $B = b / N = \frac{N-a}{N}$.

$P_a + P_b = 1$ or $P_b = 1 - P$

15.3 Steps in Testing Statistical Hypothesis

Testing of a hypothesis is a decision-making process about the population/population parameter based on evidence given by the sample drawn from the population. The steps are shown in Figure 15.4.

Test Statistic: The formula or test statistic for testing the hypothesis related to mean or proportion is:

$$\text{Test statistic} = \frac{\text{Related sample statistic} - \text{Hypothesised population parameter}}{\text{Standard error of the sample statistic}}$$

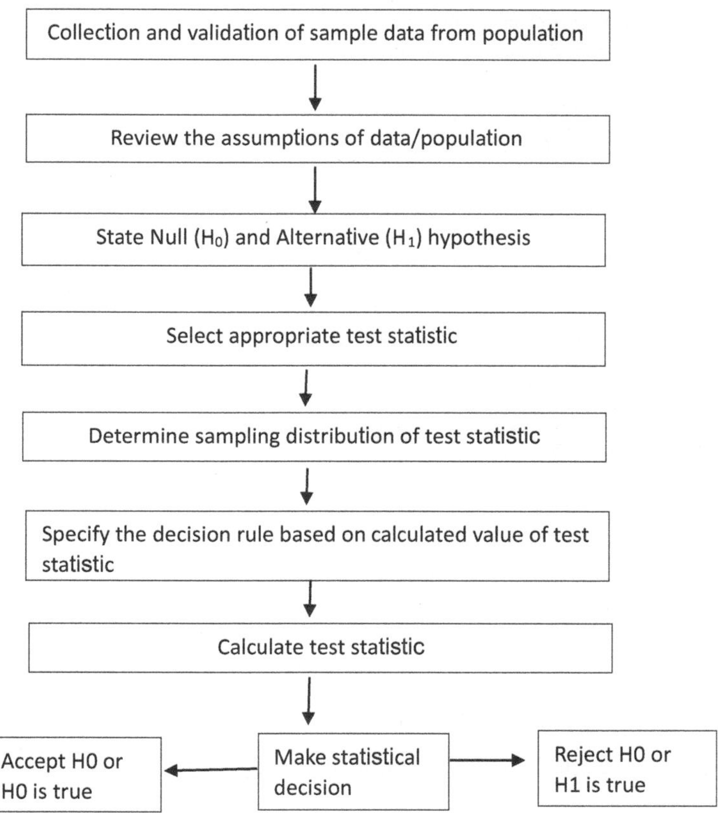

Figure 15.4 Steps in Testing Statistical Hypothesis

15.4 Point and Interval Estimation

The researcher will have a target population for his/her research and the researcher normally will have a focused outcome variable which is also called the dependent variable of the study pertaining to the population. Normally, the in-depth study of the dependent variable is based on a random sample drawn from the population. Generally, the mean value or proportion of an attribute in the population is the focus of the study which is estimated based on a representative sample.

The process of working out population parameters using random sampling technique is called estimation. There are two ways of estimating population parameters based on sample estimates—point estimate and interval estimate.

15.4.1 Point Estimation

It is getting an unbiased estimate of population parameter as a single value. When we estimate population mean μ as sample mean $\bar{X}$, the value of $\bar{X}$ is a point estimate of population mean. For large sample $E(\bar{X}) = \mu$.

15.4.2 Interval Estimation

To overcome the error or likely deviation of point estimate from population parameter, an interval is formed with point estimate at the centre and an interval with a lower limit and upper limit on both sides based on the standard error and tabled value of corresponding test statistic (t or Z) so that there is a fixed probability for the population parameter to lie within that limit. Let c_1 and c_2 are the limits of the interval then, $\{c_1<\mu<c_2\}$ is the confidence interval for μ.

If 't' is the statistic used to estimate μ, then $(1 - \alpha)$ per cent confidence interval for μ is given by $[t \pm SE(t)\, z_\alpha]$.

15.4.2.1 Confidence Interval for Population Mean (μ)

Here population mean μ is estimated as $\bar{X}$.

When X follows $N(\mu,\sigma)$ we know that $\bar{X}$ will follow $N\left(\mu,\sigma/\sqrt{n}\right)$.

Hence $\dfrac{\bar{X}-\mu}{\sigma/\sqrt{n}}$ will follow N(0,1).

We know that probability $(-1.96 \le Z \le 1.96) = 0.95$

or probability $\left(\bar{X} - 1.96\sigma/\sqrt{n} \le \mu \le \bar{X} + 1.96\,\sigma/\sqrt{n}\right) = 0.95$,

that is, 95% confidence interval for population mean $\mu = \bar{X} \pm 1.96\,\sigma/\sqrt{n}$

and 99% confidence interval for population mean $\mu = \bar{X} \pm 2.58\,\sigma/\sqrt{n}$.

When σ is not known the standard deviation of the sample (s or S) is used to make the confidence interval.

15.4.2.2 Confidence Interval for Population Proportion (P)

Here population proportion P is estimated as sample proportion p.

Hence $Z = \dfrac{p - E(p)}{S.E(p)}$ will follow N(0,1),

that is, $Z = \frac{p-P}{\sqrt{PQ/n}}$ will follow N(0,1).

Hence 95% confidence interval for $P = p \pm 1.96\sqrt{pq}/n$

and 99% confidence interval for $P = p \pm 2.58\sqrt{pq}/n$.

15.4.2.3 Confidence Interval for Difference of Means ($\mu_1 - \mu_2$)

$\bar{X}_1$ and $\bar{X}_2$ are the sample means and s_1 and s_2 are sample standard deviations based on two independent samples from two normal populations with means μ_1 and μ_2 and standard deviations σ_1 and σ_2.

95% confidence interval for $\mu_1 - \mu_2 = \bar{X}_1 - \bar{X}_2 \pm 1.96\sqrt{\frac{s_1^2}{n_1} + \frac{s_2^2}{n_2}}$

99% confidence interval for $\mu_1 - \mu_2 = \bar{X}_1 - \bar{X}_2 \pm 2.58\sqrt{\frac{s_1^2}{n_1} + \frac{s_2^2}{n_2}}$.

15.4.2.4 Confidence Interval for Difference of Proportions

Let p_1 and p_2 are sample proportions based on two independent samples from two populations with proportions P_1 and P_2.

Hence 95% confidence interval for $P_1 - P_2 = p_1 - p_2 \pm 1.96\sqrt{\frac{p_1 q_1}{n_1} + \frac{p_2 q_2}{n_2}}$

And 99% confidence interval for $P_1 - P_2 = p_1 - p_2 \pm 2.58 \pm \sqrt{\frac{p_1 q_1}{n_1} + \frac{p_2 q_2}{n_2}}$

15.5 Steps in Testing of Hypothesis

(i) Set up Null Hypothesis (H_0).
(ii) Set up Alternative Hypothesis (H_1) to decide use of single-tailed or two-tailed test.
(iii) Choose appropriate level of significance (a) or probability of Type-I error.
(iv) Compute test statistic/test criterion (SND, t, X^2, F etc.).
(v) Conclusion:

If calculated test statistic < tab value of test statistic, then accept H_0.
If calculated test statistic > tab value of test statistic, then reject H_0.

15.6 Parametric Tests

Parametric tests can be a large sample test (sample size n > 30) or a small sample test (n < 30). The various parametric tests used for testing of hypothesis are:

(i) Standard normal distribution test (SND test)
(ii) Student's t-test
(iii) Pearson's Chi-square test for goodness-of-fit
(iv) F-test

15.6.1 Standard Normal Distribution Test (SND/Z Test)

- When X → follows $N(\mu,\sigma)$, then $Z = \dfrac{X-\mu}{\sigma}$ follows N (0, 1).
- It is used for a large sample (n > 30).
- Used to test significance of single mean, single proportion, difference of two proportions, difference of two means, difference of two standard deviations.
- The test statistic $Z = \dfrac{t-E(t)}{SE(t)} \sim N(0,1)$, where 't' is the sample estimate.

SND test for single mean:

(i) **Sample is drawn from a normally distributed population with known variance (σ^2)**

$H_0 : \mu = \mu_0$ and let $H_1 : \mu \neq \mu_0$. Here the test statistic is $Z = \dfrac{(\bar{X}-\mu_0)}{\sigma/\sqrt{n}}$.

(ii) **Sample is drawn from a normally distributed population with unknown variance**

$H_0 : \mu = \mu_0$ and let $H_1 : \mu \neq \mu_0$ Here the test statistic is $Z = \dfrac{(\bar{X}-\mu_0)}{s/\sqrt{n}}$ where $s^2 = 1/n\Sigma(X-\bar{X})^2$

SND test for difference between two population means:

(i) **Sampling from normally distributed populations with known variances**

$H_0 : \mu_1 = \mu_2$; $H_1 : \mu_1 \neq \mu_2$. Here test statistic is $Z = \dfrac{(\bar{X}_1-\bar{X}_2)-(\mu_1-\mu_2)}{\sqrt{\left(\dfrac{\sigma 1^2}{n_1}+\dfrac{\sigma 2^2}{n_2}\right)}}$.

(ii) **Sampling from normally distributed populations with unknown but equal variances**

$H_0 : \mu_1 = \mu_2$; $H_1 : \mu_1 \neq \mu_2$. Here test statistic is $Z = \dfrac{(\bar{X}_1-\bar{X}_2)-(\mu_1-\mu_2)}{\sqrt{\dfrac{s_p^2}{n_1}+\dfrac{s_p^2}{n_2}}}$.

Where $s_p^2 = \dfrac{(n_1-1)s_1^2+(n_2-1)s_2^2}{n_1+n_2-2}$, here s_p^2 is the pooled variance.

(iii) **Sampling from normally distributed populations with unknown and unequal variances**

$H_0 : \mu_1 = \mu_2$; $H_1 : \mu_1 \neq \mu_2$. Here test statistic is $Z = \dfrac{(\bar{X}_1-\bar{X}_2)-(\mu_1-\mu_2)}{\sqrt{\dfrac{s_1^2}{n_1}+\dfrac{s_2^2}{n_2}}}$.

15.6.2 Student's t-Test

- If $X_1, X_2, \ldots X_n$ is a random sample of size n which follows N (μ, σ), then the student's t is defined as $t = \frac{\bar{X} - \mu}{S/\sqrt{n}}$ follows a student's distribution with n − 1 degrees of freedom.
- Where $S^2 = \frac{1}{n-1}\Sigma\left(X_i - \bar{X}\right)^2$.
- It is used for small sample (n < 30), population from which sample drawn is normal, sample observations are random, population variance is known or suitably estimated using sample variance.
- Used to test significance of single mean, difference of two means, difference of two standard deviations.
- Also used to test significance of sample correlation coefficient, estimated sample regression coefficients.
- Paired t-test is used when before and after means of the same sample are to be tested for equality.

15.6.3 Chi-Square Test (χ^2)

- If X follows $N(\mu, \sigma)$ then, $\left(\frac{X-\mu}{\sigma}\right)^2$ follows a χ^2 distribution with one degree of freedom.
- The sum of squares of n independent SND, follows chi-square with 'n' degrees of freedom.
- The total frequency (aggregate cell counts) should be reasonably large, observations should be independent (no repetition of items), no cell frequency is small (< 5).
- Used for test-for-goodness of fit, test for independence of attributes and test for homogeneity of variance (population has a specified value of variance).

15.6.4 F-Test (ANOVA)

- If X follows χ^2 with n_1 degrees of freedom and Y follows χ^2 with n_2 degrees of freedom, then F statistic is defined as $F = \frac{X/n_1}{Y/n_2}$. The G.W. Snedecor's F follows with (n_1 and n_2) DF.
- Used to test equality of two population variances.
- Test significance of an observed sample multiple correlation.
- Test significance of observed sample correlation ratio.
- Test linearity of regression.
- Test equality of three or more population means.

15.6.5 Selection of Statistical Tests

The selection of statistical tests depends on many factors like the nature and type of parental distribution (as normal or non-normal), the sample as random or purposive, the type of parameters being tested as mean values or proportions, the study hypothesis and other assumptions being made about the population under study by the researcher. The available statistical tests broadly classified as parametric and non-parametric tests. The summary of commonly used parametric tests are given in Table 15.2.

Table 15.2 Common Parametric Test Statistic for Different Situations/Purposes

S. No.	*Situation/Purpose*	*Null Hypothesis H_0:*	*Conditions*	*Test Statistic*
1.	Testing of single population mean (already available value, say, μ_0), population variance σ^2 known)	$\mu = \mu_0$	Sample drawn from N (μ, σ); SD (σ) known	SND-test/Z-test
2.	Testing of single population mean (already available value, say, μ_0), population variance (σ^2 unknown)	$\mu = \mu_0$	Sample drawn from N (μ, σ); SD (σ) unknown sample size is small (< 30)	Student's t-test
3.	Testing of single population mean (already available value, say, μ_0), population variance (σ^2 known/ unknown)		Sample drawn from non- normal population and sample size is large (n > 30)	SND-test/Z-test
4.	Testing difference between two population mean, variances known	$\mu_1 = \mu_2$ or $\mu_1 - \mu_2 = 0$	Independent samples drawn from N (μ_1, σ_1) and N (μ_2, σ_2); known population variances (σ_1^2 and σ_2^2)	SND-test/Z-test
5.	Testing difference between two population mean, variances unknown but equal	$\mu_1 = \mu_2$ or $\mu_1 - \mu_2 = 0$	Independent samples drawn from N (μ_1, σ_1) and N (μ_2, σ_2); unknown but equal population variances	Student's t-test
6.	Testing difference between two population mean, variances unknown and unequal	$\mu_1 = \mu_2$ or $\mu_1 - \mu_2 = 0$	Independent samples drawn from N (μ_1, σ_1) and N (μ_2, σ_2); unknown and unequal population variances	Student's t-test
7.	Testing difference between two population mean, sample from non-normal distribution	$\mu_1 = \mu_2$ or $\mu_1 - \mu_2 = 0$	Sample from non-normal distribution —Variances known —Variances unknown	SND-test/Z-test
8.	Testing difference between means of two non-independent samples (same sample at two points of time—paired t-test)	$\mu_d = 0$	Samples are not independent, same sample and data at two points of time taken (before and after)	Paired t-test
9.	Testing single population proportion	$P = P_0$	Sample is large, p + q = 1 or q = 1 − p	SND-test/Z-test

(*Continued*)

Table 15.2 (Continued)

S. No.	*Situation/Purpose*	*Null Hypothesis H_0:*	*Conditions*	*Test Statistic*
10.	Testing difference between two population proportions (x_1 and x_2 are number of cases possessing the characteristics) or $p_1 = x_1/n_1$ and $p_2 = x_2/n_2$	$P_1 = P_2$ or $P_1 - P_2 = 0$	When, $p_1 = p_2$, $\bar{p} = \frac{x_1 + x_2}{n_1 + n_2}$ sample is large	SND-test/Z-test
11.	Testing single population variance	$\sigma^2 = \sigma_0{}^2$	Sample is random from N (μ, σ)	Chi-square test
12.	Testing ratio of two population variance	$\sigma_1{}^2 = \sigma_2{}^2$	Samples from two normal populations	F-test
13.	Testing equality of more than three population means	$\mu_1 = \mu_2 = \mu_3 = \ldots = \mu_k$	Samples from three or more normal populations	F-test
14.	Testing significance of coefficient of determination in regression analysis	$R^2 = 0$	Sample data of both dependent and independent variable available for all units	F-test
15.	Testing significance of correlation coefficient	r = 0	Sample is from N (μ, σ) Paired data of X and Y available for all units	Student's t-test
16.	Testing significance of regression of estimates regression coefficient $\hat{Y} = \hat{a} + \hat{b}X$	$\hat{a} = 0$ $\hat{b} = 0$	Data for X and Y available for all sample units	$t_1 = \hat{a} / se(\hat{a})$ t_{n-2} $t_2 = \hat{b} / se(\hat{b})$ Student's t-test
17.	Matching of observed frequencies with theoretical frequencies (goodness-of-fit test)	Observed frequencies = Theoretical frequencies	Observed sample frequencies and basis to calculate theoretical frequency given	Chi-square test

15.7 Non-Parametric Statistical Tests

15.7.1 Introduction

The parametric tests make assumptions about the parameters of the population distribution(s) from which the sample data are drawn. The non-parametric tests do not make any such assumptions about parental population. However, relatively non-parametric statistical tests are less powerful compared to parametric tests. For example, a parametric correlation uses information about the mean and deviation from the mean while a non-parametric correlation will use only the ordinal position of pairs of scores. When tests are not concerned with population parameters or when we do not have knowledge about sampled populations (distribution-free), then we apply non-parametric tests. Non-parametric and distribution-free tests are the same. When the data is in the form of ranks or frequencies then also non-parametric tests can be used.

When to use non-parametric tests

- The inferences drawn from the parametric tests such as Z, *t* and F may be seriously affected when the parent population's distribution is not normal.
- The adverse effect could be more when sample size is small.
- Thus, whenever there is doubt about the distribution of the parent population, the non-parametric tests should be used.
- In many situations, particularly in social and behavioral sciences, observations are difficult or impossible to take on numerical scales and a suitable non-parametric test based on ordinal values is an alternative under such situations.

The assumptions about parametric tests include:

- The observations are independent.
- The observations are drawn from specifically distributed populations.
- These populations under test are having the same variances.

Certain assumptions are made with most non-parametric statistical tests, but these are fewer and weaker than those of parametric tests.

Advantages of non-parametric tests:

- Probability statements obtained from non-parametric statistics are exact probabilities, irrespective of the shape of the population distribution.
- For very small sample cases, there is no alternative to non-parametric tests.
- Non-parametric tests are comparatively easier to learn and apply than parametric tests.
- No assumption on specific form of the distribution from which the sample is drawn.
- Hence non-parametric tests are also known as distribution-free tests.

Disadvantages of non-parametric tests:

- The non-parametric tests do not fully utilize all information of the data,
- Less powerful,
- Non-availability of user-friendly software,
- Limited use of tabled value, etc.

The summary of commonly used non-parametric tests according to the number of samples, population parameter tested, null and alternative hypotheses are given in Table 15.3.

While applying non-parametric tests, it must be kept in mind that these tests are less powerful compared to parametric tests. However, when either condition to apply parametric tests is not fulfilled or the nature of data does not permit the application of parametric tests, then the option left is non-parametric tests.

The situation-wise non-parametric tests based on a number of samples dealt with by the researcher and the nature of the sample as dependent or independent are listed below.

- **Relationship between variables:** Chi-square test, Spearman rank correlation test, binomial test, run test, one sample Kolmogrov-Smirnov test.
- **Test for difference between groups**

Table 15.3 Summary of Commonly Used Non-Parametric Tests

S.	*Name of Test*	*No. of sample*	*Population Parameter Tested/ Situation/Assumptions*	*Null and Alternative Hypotheses*	*Test Criterion/Test Statistic*
1.	Sign Test (one set of ordinal data)	01	Median of rank of population. Used when the normality assumptions of *t*-test not fulfilled or the data is in rank form.	H_0: $M = M_0$ H_1: $M \neq M_0$ $M > M_0$ $M < M_0$	Number of +ve and −ve signs for deviations of observed data from median = M_0. The test statistic is sufficiently small no. of observed +ve or −ve signs or both according to the nature of alternative hypothesis. p-value calculated using binomial distribution for x number of −ve or +ve signs in a sample of size n.
2.	Sign Test (paired data)	02	Median of rank of X and rank of Y. Used when the normality assumptions of t-test not fulfilled or the data is in rank form for matched pairs.	H_0: $Mx = My$ $Mx - My = 0$ $P(Xi > Yi) = P(Xi < Yi) = 0.5$ H_1: $Mx \neq My$	Every matched score for Y is subtracted from that of X and the signs + or − given to each pair. For matched pairs, the null hypothesis becomes H_0: $P(+) = P(-) = 0.5$. If there are more numbers of +ve or −ve signs the truth of null hypothesis is doubted. The test statistic is the dominance of one sign over the other in the difference of score. The p-value calculated using binomial distribution.
3.	Wilcoxon Signed-Rank Test	01	Mean of continuous or interval data. When Z-test or *t*-test is not applicable due to small sample size or non-normality in data. The randomness and symmetric distribution of data about mean are assumed.	H_0: $\mu = \mu_0$; $\mu > \mu_0$; $\mu < \mu_0$ H_1: $\mu \neq \mu_0$; $\mu < \mu_0$ $\mu > \mu_0$	Find $d_i = x_i - \mu_0$. If any di = 0, remove that case from the sample and reduce sample size accordingly. Rank each di based on absolute value of di by ignoring the sign. For tied ranks, use averages of those ranks. Then assign a sign of actual di to each rank. Find T+ and T- as sum of ranks with +ve and −ve signs. The test statistic will be T+ or T—depending upon the nature of alternative hypothesis.
4.	Wilcoxon Matched-Pairs Signed-Rank Test	02	Mean difference of paired data.	H_0: $\mu_D = 0$ H_1: $\mu_D \neq 0$ $\mu_D > 0$ $\mu_D < 0$	Find $d_i = x_i - yi$. If any di =0, remove that case from sample and reduce sample size accordingly rank $\lvert di \rvert$. Assign a sign of di to each rank. Find T+ and T- as sum of ranks with +ve and −ve signs. The test statistic will be T+ or T- depending upon the nature of alternative hypothesis.
5.	Median Test	02	Two independent samples are drawn from a population with equal median. It is assumed that samples are random and drawn independently from respective populations. The data is at least on ordinal scale. The sample need not be of equal size.	H_0: $Mx = My$ H_1: $Mx \neq My$	Count the number of cases in X and Y which are above and below the given medians. Then form a 2 × 2 contingency table for X and Y with a number of cases above and below the medians. The test statistic will be Chi-square calculated from the 2 × 2 contingency table.

(Continued)

Table 15.3 (Continued)

S.	*Name of Test*	*No. of sample*	*Population Parameter Tested/ Situation/Assumptions*	*Null and Alternative Hypotheses*	*Test Criterion/Test Statistic*
6.	Mann-Whitney Test	02	Two populations from which the samples are drawn have equal medians	H_0: Mx = My H_1: Mx ≠ My Mx > My Mx < My	Combine the two samples and rank observations from smallest to largest keeping identity of X and Y populations. Tied observations are given equal ranks as mean of the rank's positions. The test statistic is $U_x = n_x n_y + \frac{n_x(n_x+1)}{2} - R_x$; $U_y = n_x n_y + \frac{n_y(n_y+1)}{2} - R_y$ Where, n_x and n_y are the sample size of the population X and Y. R_x and R_y are the sum of ranks of X and Y population when ranking is done for the joint data keeping the identity of both X and Y data. Out of U_x and U_y, smaller value of U is the test statistic
7.	Kolmogrov-Smirnov Goodness-of-Fit Test	01	Sample cumulative distribution $F_s(x)$ is compared with some theoretical cumulative distribution $F_t(x)$. Assumptions: (i) Sample is random. (ii) $F_t(x)$ is continuous.	H_0: $F(x) = F_t(x)$ H_1: $F(x) \neq F_t(x)$	$D = Sup_x \mid Fs(x) - Ft(x) \mid$ where D is the greatest of all absolute difference over all x between $F_s(x)$ and $F_t(x)$. Decision to accept or reject H_0 is taken for tabled value available for different values of n.
8.	Kruskal-Wallis One-Way ANOVA by Ranks	Three or more	Used when assumptions on normality and equal variances of populations are not fulfilled or observations are in rank form/ordinal scale. Assumed that all samples are independent.	H_0: The medians of all populations are equal. H_1: At least one population has greater values compared to another population.	Median of the combined sample ($n = n_1 + n_2 + \ldots + n_k$) worked out. Original values of each sample replaced with corresponding ranks. Assigned ranks in each group are added to get k sums. The test statistic is: $H = \frac{12}{n(n+1)} \sum_{j=1}^{k} \frac{R_j^2}{n_j} - 3(n+1)$ where k = number of samples; n_j = number of observations in j^{th} sample; R_j = sum of ranks in j^{th} sample; n = number of observations in all samples. The calculated H is compared with tabled Chi-square with k−1 DF. Decision to accept or reject H_0 is taken for tabled value.
9.	Friedman Two-Way ANOVA by Ranks	Three or more treatments	Observations in each block are independent measurements in ordinal scale.	H_0: All treatments are identical. H_1: At least one treatment has greater values compared to another treatment.	Assume that the ranks occur equally in its frequencies if H_0 is true, otherwise at least for one treatment dominance of high or low is expected which can be reflected in sum of ranks for different treatments. The test statistic derived by Friedman is blocks; k = number of treatments; Rj = sum of ranks of jth treatment. Decision to accept or reject H_0 is taken for tabled value.

(i) **Independent Samples—two variables:** Mann-Whitney U test, Kolmogrov-Smirnov two-sample test, Wald-Wolfowitz run, Moses extreme reactions.

(ii) **Independent Samples—more than two variables:** Median test, Kruskal-Wallis H test.

(iii) **Dependent Sample—two-variable:** Wilcoxon signed-rank, McNemar test, marginal homogeneity, sign test.

(iv) **Dependent Sample—more than two variables:** Cochran test, Friedman test.

Some Important Points

- **Statistical Inference** is the process of making a decision about population (parameters) based on evidence given by a sample (statistic). It has two parts, first, it is an estimation of population parameters based on sample data and second it is testing the significance of sample estimates for population parameters.
- **Statistical Hypothesis:** A statistical hypothesis is a statement about population (may or may not be true) which is tested based on evidence from sample using the theory of probability.
- **Null Hypothesis (H_0):** Null hypothesis is the hypothesis which is tested under the assumption that it is true. It is generally the hypothesis of no difference. It is drawn based on a neutral or null attitude by the decision maker.
- **Alternative Hypothesis (H_1):** It is that hypothesis which is accepted if the null hypothesis is rejected.
- **Simple Hypothesis:** If the hypothesis completely specifies the population, it is simple.
- **Composite Hypothesis:** If the hypothesis does not completely specify the population, then it is a composite hypothesis: $H_1 : \mu < \mu_0; \mu > \mu_0; \mu \neq \mu_0$.
- **Test of Significance:** It is the procedure to assess the significance of a statistic or difference between two independent statistics.
- **Types of Errors:** When we make inference about population parameters based on evidence given by the sample, two types of errors are possible. First, the sample evidence may lead to reject a null hypothesis, when it is actually true. It is called Type-I error. Second, the sample evidence may lead to accept a null hypothesis, when it is actually false. It is called Type-II error.
- **Test Criteria** fix the size of Type-I error at a low value and devise test criteria which minimize Type-II error.
- **Power of Test** = 1 − Prob. (Type-II error).
- **Level of Significance:** The size of Type-I error, that is, Pr/Rejecting H_0/H_0 is true. It is the probability of rejecting a true null hypothesis. It is generally kept at 5% (0.05) or 1% (0.01).
- Confidence coefficient (%) = 1 − Type-I error
- **Critical Value or Significant Value:** The value of test statistic which separates the critical region and acceptance region is called critical value or significant value.
- **One-sided and two-sided tests:** Depending upon the nature of the alternative hypothesis, a test can be one-sided or two-sided. If the alternative hypothesis is of the type 'not equal to', then the alternative hypothesis can fall on either side of the curve as given below and if it is of the type 'greater than or less than' then it falls either on the right side or on the left side of the curve.
- **Sampling Distribution** is the set of values of the test statistic. Like any other variable, it can also have mean and variance
- **Standard Error (SE)** is the standard deviation of the sampling distribution of a statistic SE(t).

- **Precision:** The precision of a sample estimate for a population parameter is the reciprocal of the SE of the estimate.
- **Estimation:** The process of working out population parameters using random sampling technique is called estimation. There are two ways of estimating population parameters based on sample estimates—point estimates and interval estimates.
- **Point Estimation:** It is getting an unbiased estimate of population parameters as a single value.
- **Interval Estimation:** In order to overcome the likely error or deviation of point estimate from the population parameter, an interval is formed with a point estimate at the centre and an interval with a lower limit and upper limit on both sides based on the standard error and tabled value of corresponding test statistic (t or Z) so that there is a fixed probability for the population parameter to lie within that limit. Let C1 and C2 be the limits of the interval then, $\{C1 < \mu < C2\}$ is the confidence interval.
- **Steps in Testing of Hypothesis** include: (i) Set up null hypothesis (H_0); (ii) set up alternative hypothesis (H_1) to decide use of single-tailed or two-tailed test; (iii) choose appropriate level of significance or probability of Type-I error; (iv) compute test statistic/test criterion (SND, t, X^2, F, etc.); (v) conclusion as if calculated test statistic < tab value of test statistic, then accept H_0 and if calculated test statistic > tab test statistic, reject H_0.
- **Parametric Tests** can be a large sample test (sample size n > 30) or a small sample test (n < 30). The various parametric tests used for testing the hypothesis are standard normal test (SND test), Student's t-test, Pearson's chi-square test for goodness-of-fit and F-test.
- **The selection of statistical tests** depends on many factors like the nature and type of parental distribution (as normal or non-normal), the sample as random or purposive, the type of parameter being tested as mean values or proportions, the study hypothesis and other assumptions being made about the population under study by the researcher. The available statistical tests are broadly classified as parametric and non-parametric tests.
- **Non-Parametric Tests**: The non-parametric tests do not make any assumptions about parental population. A non-parametric correlation will use only the ordinal position of pairs of scores. When tests are not concerned with population parameters, or when we do not have knowledge about the sampled population (distribution-free), then we apply non-parametric tests. Non-parametric and distribution-free tests are the same. When data is in the form of ranks or frequencies, then also non-parametric tests can be used.
- **Non-Parametric Tests** are available for the relationship between variables and for testing the difference between independent and dependent groups.

Suggested Readings

Altman, D. G., *Practical statistics for medical research*, Chapman & Hall, London, 1992.

Brown, L. D., T. Cai, and A. DasGupta, 'Interval estimation in exponential families', *Statistica Sinica* 13: 19–49, 2001.

Brown, L. D., T. Cai, and A. DasGupta, 'Confidence intervals for a binomial proportion and asymptotic expansions', *The Annals of Statistics* 30(4): 160–201, 2002.

Daniel, W. W., *Biostatistics: basic concepts and methodology for the health sciences*, Wiley Publications, John Wiley & Sons, Reprint by Wiley India (P) Ltd., New Delhi, 2010.

Dunn, O. J., *Basic statistics*, Wiley, New York, 1984.

Gupta, S. C., *Fundamentals of statistics*, Himalaya Publishing House, Mumbai, 1981.

Lehmann, E. L., *Nonparametrics: statistical methods based on ranks*, McGraw-Hill, San Francisco, CA, 1985.

Negi, K. S., *Biostatistics*, 2nd ed., AITBS, New Delhi, 2010.

Patel, J. K., and C. B. Read, *Handbook of the normal distribution*, Marcel Dekker, New York, 1982.
Rao, C. R., *Linear statistical inference and its applications*, 2nd ed., John Wiley & Sons, New York, 1973.
Siegel, S., and N. J. Castellan, *Nonparametric statistics for the behavioral sciences*, 2nd ed., McGraw-Hill, New York, 1988.
Sprent, P., *Applied nonparametric statistical methods*, Chapman & Hall, Losu, 1989.
Zar, J. H., *Bio-statistical analysis*, 4th ed., Prentice-Hall, Upper Saddle River, NJ, 1999.

16 Multivariate Statistical Techniques

16.1 Introduction

In most of the univariate analysis, one variable which follows a normal population is generally assumed. When many variables or measurements in each unit of study are simultaneously considered, it forms a case for multivariate analysis. In agriculture, a farming system study may include variables related to farm family (family size, number of male members/female members, children, etc.); farm animals (number of cows, buffaloes, goats, draught animals, etc.); farm area (farm area, irrigated area, cropped area, etc.), farm household economy (income, consumption expenditure, family savings, etc.) and so on. Each of these variables may follow a normal distribution individually. In medical and health research, a set of observations of a patient related to anthropometry (age, height, weight, etc.); haematology (blood pressure, haemoglobin counts, platelet counts, etc.); other biochemical, pathological, microbiological, etc. are considered for each patient as components of the random vector. Instead of a random variable in univariate approach, a random vector consisting of a column matrix is the basis of multivariate approach. Research in other areas like business, marketing, sociology, psychology, economics and others where a set of observations are considered as a random vector on each unit of study, it is possible to have multivariate analysis.

In multivariate analysis, a set of measurements on each unit of the study is taken and treated as a vector. When similar observations are available for a given number of units of study and all these observations are simultaneously considered for analysis, it forms the case of Multivariate analysis. In the previously stated case of multivariate analysis, the given set of data of an agricultural household or a patient is treated as a vector (a column matrix). If the vector has all the characteristics of a random variable and it is called a random vector and denoted as $[X] = [X_1, X_2, X_3, \ldots, X_p]^T$, where T standard for transpose of the column matrix. Similar to normal distribution, in univariate analysis in one variable case, there is multivariate normal distribution in the multivariate case as:

$$f(x) = \frac{|\Sigma|^{\frac{-1}{2}}}{(2\pi)^{\frac{p}{2}}} e^{-\frac{1}{2}(X-\mu)^{|}\Sigma^{-1}(X-\mu)}.$$

Here, μ is the mean vector of individual variables $X_1, X_2, \ldots, X_p$ and Σ is the variance-covariance matrix of the components of the vector [X]. Thus μ is a column matrix with mean values of the individual components and Σ is a 'p x p' square matrix with diagonal elements as the variances of $X_1, X_2, \ldots, X_p$ and non-diagonal elements are co-variances of X_i and X_j, $I \neq j$.

DOI: 10.4324/9781003527183-16

If there are p components in the vector variable X, then $X = [X_1, X_2, \ldots, X_p]^T$ is the multivariate random vector with population mean vector as $\mu = [\mu_1, \mu_2, \ldots, \mu_p]^T$ and population variance-covariance matrix Σ is:

$$\Sigma = \begin{bmatrix} \sigma_{11} & \sigma_{12} & \sigma_{1p} \\ \ldots & \ldots & \ldots \\ \sigma_{p1} & \sigma_{p2} & \sigma_{pp} \end{bmatrix}.$$

The corresponding sample vector for mean

$\bar{X} = \left[\overline{X_1}, \overline{X_2}, \ldots \overline{X_p}\right]^T$ and sample variance-covariance matrix is $S = \begin{bmatrix} S_{11} & S_{12} & S_{1p} \\ \ldots & \ldots & \ldots \\ S_{p1} & S_{p2} & S_{pp} \end{bmatrix}$.

The variance-covariance matrix is a symmetric matrix meaning all $S_{ij} = S_{ji}$. In other words, $S_{12} = S_{21}$, $S_{13} = S_{31}$. When households from rural and urban areas or patients with and without disease are to be compared using multivariate analysis, two separate multivariate normal distributions are assumed.

The multivariate analysis has a wide range of applications in research in areas like agriculture, health and medical sciences, business, commerce, economics, social sciences and many others. The major type of multivariate analysis includes the following:

(i) Testing equality of mean vectors for two populations.
(ii) Discriminant analysis.
(iii) Multiple regression analysis.
(iv) Principal component analysis.
(v) Factor analysis.
(vi) Cluster analysis.
(vii) Canonical correlation.
(viii) Structural equation modelling, etc.

16.2 Testing Equality of Mean Vectors for Two Populations—Hotelling's T^2

There may be multivariate cases like irrigated farming systems versus unirrigated farming systems; rural households versus urban households; normal individuals versus sick individuals; traditional methods versus technological methods where a set of similar measurements may be available for both the cases or populations under study.

The researcher may be interested to test equality of mean vectors (μ_x) and (μ_y) of two multivariate normal distributions. The equality of variance-covariance matrix/homogeneity of variance is assumed for both populations.

The null hypothesis is: $H_0 : (\mu_x) = (\mu_y)$.

The alternative hypothesis is: $H_1 : (\mu_x) \neq (\mu_y)$.

The test statistic is F as defined:

$$F(p, n_1 + n_2 + -p - 1) = \frac{n_1 n_2}{n_1 + n_2} \frac{n_1 + n_2 - p - 1}{p} (\bar{X} - \bar{Y})^T S^{-1} (\bar{X} - \bar{Y}).$$

Where n_1 = size of sample from Population I; n_2 size of sample from Population II
$\bar{X}$ = mean vector of sample from Population I; $\bar{Y}$ = mean vector of sample from Population II
S^{-1} = the inverse of variance-covariance matrix of pooled sample from both the populations
P = number of components in random vectors X and Y of Populations I and II respectively.

The decision to accept or reject the null hypothesis is taken by comparing calculated F value with tabled F value corresponding to given degrees of freedom $(p, n_1 + n_2 - P - 1)$.

16.3 Discriminant Analysis

It is a multivariate technique used to classify units of a multivariate population in defined classes based on available multivariate measurements on subjects fallen in these defined classes of the said population. There should be at least two classes to classify the units of the population under study as passing or failing an examination; curing or not curing from a disease after treatment; profit earning or loss incurring in a business and so on. For simplicity let us assume a two-group class situation for which the dataset is available separately. These two classes are assumed as separate populations. The discriminant function and the rule for discrimination can be worked out based on the mean vectors and variance-covariance matrices for the defined classes. Once the function and the rule for classification are fixed, any new subject from the joint population can be assessed in advance so as to fall into one of the classes prior to the actual realization of the outcome after treatment and complex calculations.

Let us assume that:

(i) (X_1) and (X_2) are the random vectors representing the units from the two groups as two separate multivariate normal distributions with p common components of measurements.

Group I as (X_1): $(x_{11}, x_{12}, \ldots x_{1p})^T$ with population mean $(\mu_1) = (\mu_{11}, \mu_{12}, \ldots, \mu_{1p})^T$
Group II as (X_2): $(x_{21}, x_{22}, \ldots, x_{2p})^T$ with population mean $(\mu_2) = (\mu_{21}, \mu_{22}, \ldots, \mu_{2p})^T$

(ii) Let the size of Group I with vector (X_1) be n_1 and for Group II with vector (X_2) be n_2.

(iii) The multivariate random vectors of the populations are assumed to have a common variance-covariance matrix (σ_{ij}) which is estimated as (S_{ij}).

(iv) Compute the sample mean vectors of (X_1) and (X_2) as $(\bar{X}_1)$ and $(\bar{X}_2)$ respectively.

(v) Compute the inverse of pooled variance-covariance (S_{ij}) as (S^{ij}).

(vi) Compute the difference of the two mean vectors as $(D_j) = (\bar{X}_1 - \bar{X}_2)$.

(vii) The discriminant function is a linear function of the form $L = l_1x_1 + l_2x_2 + \ldots l_px_p$ where $l_i = \Sigma_j S^{ij} D_j$.

(viii) The grouping of the new units is made in Group I or Group II using the value of the function 'L' such that:

$\{X / L \geq C\}$ means the unit will fall in Group I
$\{X / L < C\}$ means the unit will fall in Group II

where $C = \text{In } P_1 / P_2 - \left[\left(\bar{X}_1^T S^{ij} \bar{X}_1\right) - \left(\bar{X}_2^T S^{ij} \bar{X}_2\right)\right] / 2$

$P_1 (= n_1 / n)$ and $P_2 (= n_2 / n)$ are probabilities of units falling in Group I and Group II respectively.

(ix) The statistical significance of discriminant function can be tested using F statistic as

$$F(p, n_1 + n_2 - p - 1) = \frac{n_1 + n_2 - p - 1}{p} \times \frac{n_1 . n_2}{(n_1 + n_2)(n_1 + n_2 - 2)} \times M_p^2 \text{ where } M_p^2 = \Sigma l_i D_i$$

Li and Di are as per (vi) and (vii).

(x) The discriminant function can be validated by examining the errors in classification of known cases of n_1 and n_2 respectively.

16.4 Principal Component Analysis

The researchers may come across following instances in multiple regression analysis:

- There is high linear correlation among explanatory or independent variables.
- The number of observations (sample size) may not be as large as compared to number of explanatory variables.

The principal component analysis can be a solution in such cases. Instead of trying individual variables, a set of linear combination of 'm' independent variable is used to relate the relationship between dependent and independent variables. Such linear combinations are mutually uncorrelated and will have maximum variance between them.

16.4.1 Estimation Procedure of Principal Component

Let 'A' be the correlation matrix (symmetric) of the 'm' explanatory variables and I be an identity matrix of size maximum.

The solutions to the following 'm' equation $A - \hat{b} \cdot \mathrm{I} = 0$ are called the characteristic roots or latent roots or eigenvalues where A is 'm × m' correlation matrix of explanatory variables and I is an identity matrix of size 'm × m' and 'b' is a scalar matrix of size 'm × m'. The solutions of 'm' equations will give 'm' different values of $\hat{b}$. Let $\hat{b}_1, \hat{b}_2, \ldots, \hat{b}_m$ be the characteristic roots arranged in descending order of its size. Corresponding to each characteristic root (say $\hat{b}_j$) a characteristic vector or eigenvector ($\hat{a}_j$) of size m × 1 can be obtained by solving the following equation:

$$\left[\mathrm{A} - \hat{b}_j \mathrm{I}\right] \cdot \left[\hat{\boldsymbol{a}}_j\right] = [0].$$

Where 'A' is m × m matrix as stated above, $\hat{b}_j$ is the scalar, I is an m × m identity matrix, $\hat{\boldsymbol{a}}_j$ is m × 1 column vector of unknown values corresponding to $\hat{b}_j$ and 0 is m × 1 column matrix of zero values.

Now, normalize $\hat{a}_j$, so that $\hat{a}_j' \times \hat{a}_j = 1$.

After estimating $\hat{a}_j$, the principal component corresponding to j[th] characteristic root ($\hat{b}_j$) is defined as:

$$\hat{Z}_j = \hat{a}_{12}\mathrm{X}_1 + \hat{a}_{12}\mathrm{X}_2 + \ldots + \hat{a}_{1m}\mathrm{X}_m.$$

The ratio of $\hat{b}_j$ to $\Sigma\hat{b}_j$ is the proportion of total variation in all explanatory variables which is accounted for by the jth principal component.

In practice, only a few principal components are finally selected according to their magnitude which account for most of the variability in explanatory variables.

The correlation of each of these principal components with each of the individual explanatory variables are then calculated to find out the most closely associated explanatory variables with each major principal component. These explanatory variables can be used in the regression analysis in place of all available explanatory variables. Hence principal component analysis can be used to select most contributing explanatory variables in regression analysis, especially when sample size is small, and number of explanatory variables is large.

For a regression model, $Y = f(X_1, \cdot X_2, \ldots X_m)$ when sample size 'n' is low compared to 'm' the number of independent/explanatory variable, that is, m < n or when there is high degree of multi-collinearity among X_1, X_2, . . ., X_m, a linear combination of X_1, X_2, . . ., X_m is generated which is called the principal component.

16.5 Canonical Correlation

It is a flexible multivariate technique which simultaneously correlates independent and dependent variables. Independent variables are metric such as sales, measurements, etc. It can also utilize nonmetric categorical variables, there are only very few restrictions in this method.

In multiple linear regression of type $Z = f(x_1, x_2, x_3, \ldots, x_p)$, a linear combination of the vector of explanatory variables (X) on a single dependent variable Y is established so that the multiple correlation $R_{Y.X_1X_2\ldots X_p}$ is maximum.

Canonical Correlation is the correlation between two vectors of variables $Y = (Y_1, Y_2, \ldots, Y_p)^T$ and $X = (X_1, X_2, \ldots, X_p)^T$.

Let $Z_1 = a_1Y_1 + a_2Y_2 + \ldots + a_pY_p = a'Y$

$Z_2 = b_1X_1 + b_2X_2 + \ldots + b_pX_p = b'X$

We have to estimate ai and bj such that correlation between Z_1 and Z_2 is maximum.

$$\text{Let } p = \frac{\text{Covariance}(a'Y, b'X)}{\sqrt{\text{Variance}(a'Y) \times \text{Variance}(b'X)}}.$$

We have to choose 'a' and 'b' in such a way that p is maximum and then p is called the first canonical correlation between Y and X. Here Z_1 and Z_2 are called first set of canonical variables associated with Y and X.

The canonical correlation and variables can be obtained by solving the constrained maximization problem:

$$\text{Maximize } \frac{a'Y.\, b'X}{\sqrt{(a'\Sigma 11\, a)(b'\Sigma 22 b)}} \text{ subject to } a'\Sigma_{11}a = b'\Sigma_{22}b = I_{p \times p}.$$

It can be shown that there are p canonical correlations and p canonical variables corresponding to each root of the equation: $\begin{vmatrix} -\lambda\Sigma_{11} & \Sigma_{12} \\ \Sigma_{21} & \lambda\Sigma_{22} \end{vmatrix} = 0$.

Here λ is called Lagrange multiplier.

Generally, out of these p roots, the highest one is chosen.

For estimation of canonical correlation or canonical variables, one has to estimate the variance-covariance matrix or the correlation matrix of the variables.

16.6 Multivariate Analysis of Variance (MANOVA)

It is used in experimental designs to assess the relationship between many categorical independent variables and two or more metric dependent variables.

It examines the dependence relationship between a set of dependent variables across a set of independent variables. A hypothesized relationship between the dependent variable is used. Independent variables are categorical and dependent variables are metric in nature. Normality of the dependent variable is necessary.

Cell sizes should be almost equal and the largest cell can have up to 1.5 times the observations of the smallest cell. Fifteen to 20 observations per cell is needed and more than 30 observations per cell lead are needed to lose practical significance. If there is a significant difference between the means, the set null hypothesis is rejected and treatment differences can be obtained.

16.7 Factor Analysis

Factor Analysis is a multivariate technique to reduce the 'p' dimensions of measurements of each object into a smaller number of factors using the original variance-covariance matrix.

If X is a 'p' dimensional multivariate random variable, that is, $\cdot X = \left[X_1, X_2, \ldots, X_p\right]'$ with E (X) = μ and variance-covariance matrix of $X = \Sigma$.

A linear structure for each of the random variable Xi is assumed:

$$X_i = \mu_i + a_{i1}Y_1 + a_{i2}Y_2 + \ldots + a_{im}Y_m + Z_i \text{ where } i = 1, 2, \ldots, p.$$

Here, Y_1, Y_2, . . ., Ym are hypothetical unobservable random variables which are common in linear structure for each Xi.

Z_i is a hypothetical unobservable random variable specific to each X_i.

a_{ij} (i = 1, 2, . . ., p and j = 1,2, . . ., m) are the coefficients of random variable Y_i.

All random variables Y_1, Y_2, . . ., Y_m appear in the linear model for each X_i and Y_j is the jth common factor while:

'm' is called the complexity of factor as small values of m implies less complexity;
Z_1, Z_2, . . ., Z_p are called specific or unique factors;
a_{ij} are called factor loading for X_i on common factor Y_i.

Factor analysis is a technique in which 'k' unobservable factors F_1, F_2, . . ., F_k are generated out of 'p' number of observable variables (Y_1, Y_2, . . ., Y_p) of interest in a study.

As a simple case, let there be three variables Y_1, Y_2 and Y_3 as marks obtained in three subjects (science, mathematics and English). Data of five students whose marks obtained in science, mathematics and English out of 10 (maximum marks) are available in Table 16.1.

Table 16.1 An Example of Factor Analysis

Student No.	*Y_1 (Science)*	*Y_1 (Mathematics)*	*Y_1 (English)*
1	3	6	5
2	7	3	3
3	10	9	8
4	3	9	7
5	10	6	5

These marks are functions of two factors F_1 (mathematical ability) and F_2 (verbal ability), respectively.

Here each Y variable can be assumed as linearly related to these two factors:

$$Y_1 = B_{10} + B_{11}F_1 + B_{12}F_2 + e_1$$

$$Y_2 = B_{20} + B_{21}F_1 + B_{22}F_2 + e_2$$

$$Y_3 = B_{30} + B_{31}F_1 + B_{32}F_2 + e_3$$

As the relationships are not exact the errors e_1, e_2 and e_3 are included in the above relationships.

The parameters B_{ij} appearing in the relationship are called factor loadings of the variables of study.

Based on the assumption that the error term e_i s are independent of one another and the unobservable factors F_j s are independent of each other and of the error term and using the variance covariance matrix of given data the factor loadings can be worked out.

It is used to reduce the large number of variables to a small set of factors when large number of variables is present in the study. It is an independent technique with no dependent variable. Independent variables are normal and continuous. Generally, three to five variables are loaded into a factor. Sample size should be greater than 50, with a minimum of five observations per variable. High multi-collinearity between variables as revealed from the correlation matrix of variables is assumed. Predictability of every variable by all other variables is assessed using Kaiser's measure of statistical adequacy (MSA); MSA > 0.8 is good and MSA < 0.5 is poor.

There are two methods of factor analysis: (i) Common factor analysis (CFA) and (ii) principal component analysis (PCA). CMA extracts factors based on variances shared by factors and is used to look for latent factors. PCA extracts factors based on total variance of the factors to find out a few numbers of variables that explain the maximum variance. The first factor extracted explains maximum variance and factors are extracted as long as eigenvalues are greater than 1.0. Factor loadings are the correlation between the factor and the variables. An orthogonal rotation assumes no correlation between factors and a factor loading greater than 0.4 is required to attribute a specific variable to a factor.

16.8 Cluster Analysis

Cluster Analysis is a technique by which 'n' objects under study are grouped into homogenous clusters based on similarities of 'p' measurable characteristics on each subject. Hence clustering is the grouping of objects according to homogeneity of multivariate characteristics. Clustering is empirically done based on similarity. There are a large number of methods for clustering and the objective of all methods is to form clusters in such a way that objects within a cluster are more or less homogenous and objects between clusters are heterogeneous.

The clustering problem as a multivariate statistical technique can be better stated as the grouping of 'n' objects (1, 2, . . ., n) into 'k' clusters (1, 2, . . ., k) which are homogenous when based on 'p' measurements (X_1, X_2, . . ., X_p) of each object. Given objects and measurements similarity is assessed based on a distance measure between each pair of objects, say, d(x,y) as the distance between X and Y. For the smaller value of d(x,y), the higher is the similarity.

It is used to reduce a large dataset to a meaningful subgroup of objects based on the similarity of objects across a set of specified characteristics. A major problem arises from outliers to too many irrelevant variables. The sample should be representative of the population, and it is desirable to have uncorrelated factors.

There are three clustering methods: (i) Hierarchal with tree-like process, (ii) non-hierarchal with priori specification of a number of clusters and (iii) combination of (i) and (ii).

Four rules to develop clusters—clusters should be different, reachable, measurable and profitable. (Cluster analysis is mostly done for market segmentation.)

The conceptual part of some of the major multivariate techniques is introduced here. The techniques of multivariate analysis are matrix-based and computer packages are available in SPSS and other software packages. Researchers aiming for system analysis based on multiple observations on each of the study units can apply to the multivariate techniques.

Some Important Points

- Multivariate statistical techniques are applied when multiple observations on each sample study unit are available in the form of a random vector, which is a column matrix of the observations of the study unit, that is, $[X] = [X_1, X_2, \ldots, X_p]^T$.
- Similar to a univariable normal distribution, there is a multivariate normal distribution with a mean vector (of size $p \times 1$) as well as variance covariance matrix (of size $p \times p$) for population as well as for corresponding sample. The sample mean vector and variance-covariance matrix of a random sample are considered unbiased estimates of population mean vector and variance-covariance matrix respectively.
- Like *t*-test for equality of two population mean in a univariable case, the equality of mean vectors of two multivariate normal populations can be tested using Hotelling's T^2, assuming a common variance-covariance matrix for the two multivariate populations.
- Discriminant analysis is a multivariate technique used to classify new units of a population in defined classes based on available multivariate measurements on subjects fallen in these defined classes of the said population. There should be at least two classes to classify the units of the population under study. The discriminant function and the rule for discrimination can be worked out based on the mean vectors and variance-covariance matrices for the defined classes.
- Principal component analysis is a linear combination of independent variables used to relate the relationship between dependent and independent variables. Such linear combinations are mutually uncorrelated and will have maximum variance between them. It is used when there is a high linear correlation among explanatory or independent variables, or the number of observations (sample size) may not be large as compared to a number of explanatory variables.
- Canonical Correlation is the correlation between two vectors of variables $Y = (Y_1, Y_2, \ldots, Y_p)^T$ and $X = (X_1, X_2, \ldots, X_p)^T$. Multivariate analysis of variance (MANOVA) is used in experimental design to assess the relationship between many categorical independent variables and two or more metric dependent variables. It examines the dependence relationship between a set of dependent variables across a set of groups.
- Factor analysis is a multivariate technique to reduce the 'p' dimensions of measurements of each object into a smaller number of factors using the original variance-covariance matrix.
- Cluster analysis is a technique by which 'n' objects under study are grouped into homogenous clusters based on similarities of 'p' measurable characteristics on each subject. Hence clustering is grouping of objects according to homogeneity of multivariate characteristics.

Suggested Readings

Anderberg, M. R., *Cluster analysis for applications*, Academic Press, New York, 1973.
Anderson, T. W., *Introduction to multivariate statistical analysis*, John Wiley & Sons, New York, 1958.
Bock, R. D., *Multivariate statistical methods in behavioral research*, McGraw-Hill, New York, 1975.
Dempster, A. P., *Elements of continuous multivariate analysis*, Addison-Wesley Pub. Comp., Reading, MA, 1969.
Dillon, W. R., and M. Goldstein, *Multivariate analysis: methods and applications*, John Wiley & Sons, New York, 1984.
Green, P. E., *Analyzing multivariate data*, The Dryden Press, Hinsdale, IL, 1978.
Johnson, R., and D. W. Wichern, *Applied multivariate statistical analysis*, Prentice-Hall, Englewood Cliffs, NJ, 1982.
Karson, M. J., *Multivariate statistical methods—An introduction*, The IOWA State University Press, IA, Ames, Iowa, 1982.
Mardia, K. V., J. T. Kent, and J. M. Bibby, *Multivariate Analysis*, Academic Press, New York, 1979.
Morrison, D. F., *Multivariate statistical methods*, McGraw-Hill, New York, 1976.
Seber, G. A. F., *Multivariate observations*, John Wiley & Sons, New York, 1984.

17 Some Other Quantitative Techniques in Research

17.1 Kappa Statistics

It is a measure of inter-observer/inter-method agreement on the interpretation of situations like consistency in two or more doctors' diagnoses of medical test results or inter-judgement consistency in participants' performance or similar such events. The inferences based on the findings of physical examination, radiographic interpretations or other lab tests by physicians/medical observers have some degree of subjective interpretation. Similar variation exists in judgements of two or more experts for cultural events, etc. Statistical tests are available to measure the extent of agreement between two or more such observers as sometimes the agreement or disagreement based on test results is simply by chance. The Kappa statistic or Kappa coefficient is a commonly used statistical measure for assessing the same. The Kappa coefficient lies between −1 and +1. The value +1 for the Kappa statistic means perfect agreement among observers, the Kappa statistic zero means the agreement is attributable to simply chance and the Kappa statistic −1 means complete disagreement among observers. The agreement between observers or inter-observer agreement measures the precision. Kappa statistics measure the precision and is a measure of the magnitude of agreement between observers. It also is a measure of reliability of the test.

The opinion of two international tourists on their choice for revisit to 100 cities visited in India during their first visit is summarized in Table 17.1.

Here, (a) and (d) represent the number of cities the two tourists completely agree for choice of revisit while (b) and (c) indicate the number of cities the two tourists disagree or have different opinion. If there were no disagreement, then (b) and (c) would have been zero, (then the observed agreement $p_0 = 1$) and if there were no agreement, then (b) and (c) would be zero and (then the observed agreement $p_0 = 0$).

The expected frequencies/values (calculated as in the case of chi-square) for a = 5, b = 15, c = 20 and d = 60.

The expected agreement of (a) = 5/100 = 0.05; the expected agreement of (d) = 60/100 = 0.60.

Table 17.1 Frequency of Usefulness of the Lectures Assessed by the Two Residents

Tourist II	*Tourist I*		*Total*
	Revisit	*No revisit*	
Revisit	15(a)	5(b)	20(m_1)
No revisit	10(c)	70(d)	80(m_0)
Total	25(n_1)	75(n_0)	100(n)

DOI: 10.4324/9781003527183-17

Table 17.2 Interpretation of Kappa Values

Kappa Value	*Interpretation*
K: < 0	Less than chance agreement (disagreement);
K= 0	Chance agreement;
K = 0.01 to 0.20	Slight agreements;
K = 0.21 to 0.40	Fair agreement;
K = 0.41 to 0.60	Moderate agreement;
K = 0.61 to 0.80	Good agreement;
0.81 to 0.99	Almost perfect agreement.

Total expected agreement $p_e = 0.05 + 0.60 = 0.65$.

Total observed agreement $p_o = (15 + 70)/100 = 0.85$.

The Kappa statistic $K = \frac{p_o - p_e}{1 - p_e} = \frac{0.85 - 0.65}{1 - 0.65} = 0.20/0.35 = 0.57$.

Interpretation of Kappa is found in Table 17.2.

17.2 Composite Index

The statistical procedure for estimation of composite index was developed by Prem Narain et al. to classify districts based on social development indicators which are summarized:

Let $[X_{ij}]$ denote the data matrix representing the development-related indicators of states; i = 1, 2, . . ., n states and j = 1, 2, . . ., k indicators.

As $[X_{ij}]$ denotes different indicators in different units of measurement these are not additive as such to get the required composite index. Hence the $[X_{ij}]$ are transformed to $[Z_{ij}]$:

$$[Z_{ij}] \frac{(X_{ij} - \bar{X}_j)}{S_j}$$

Where,

$\overline{X_J}$ = mean of jth indicator

S_j = standard deviation of jth indicators across states

$[Z_{ij}]$ = the matrix of standardized indicators.

From $[Z_{ij}]$, identify the best value of each indicator. In the case of positive (pushing factors) indicators, the best value can be the maximum state value and in the case of negative (pulling factors) indicators, it can be the minimum state value, depending upon the direction of the impact of the indicator on the level of development. Let Z_{0j} denote the best value (maximum for positive indicator and minimum for negative) of jth indicator. In order to get the pattern of development, calculate first P_{ij}

Where,

$$P_{ij} = (Z_{ij} - Z_{0j})^2.$$

The pattern of development C_i is given:

$$C_i = \sum_{j=1}^{k}\left[\frac{P_{ij}}{CV_j}\right]^{1/2}$$

Where,

CV_j is the coefficient of variation of jth indicator in matrix X_{ij}.

Composite index Di is given,

$$D_i = C_i / C$$

Where,

$$C = \bar{C} + 3SD$$

$$\text{Mean of } C_i = \bar{C} = \frac{\sum_{i=1}^{n} C_i}{n}$$

$$\text{Standard deviation of } C_i = SD = \sqrt{\frac{\sum_{i=1}^{n}\left(C_i - \bar{C}\right)^2}{n}}$$

Smaller value of D_i will indicate a high level of development and a higher value of D_i will indicate a low level of development as deviation of the standardized indicator from ideal standardized state value (maximum/minimum) are taken to calculate the composite indices.

17.3 Trend and Growth Rates

Trend is the pattern of the curve of any time series study variable like population, area, production, etc. in time points. The trend can be linear or non-linear.

$$Y = a + bt - \text{linear},$$

Where, Y = yearly area, production, sale, profit;

t = 1, 2, . . ., n number of years;

a = intercept and

b = slope of trend equation.

$$Y = ab^t - \text{exponential}$$

Where, Y and t are as above;

a = constant and $b = 1 + r$;

Where, r = exponential growth rate.

$$Y = a + bt + ct^2$$

Where, Y and t are as above; a, b and c are coefficients of the quadratic equation. Depending upon the sign of c is +ve or −ve, the trend curve can be U-shaped or inverted U-shaped.

$Y = a + b_1 t + b_2 t^2 + b_3 t^3 + \ldots + b_k t^k$, is a kth polynomial trend equation;

Where, $b_1, b_2, \ldots, b_k$ are the coefficients of kth polynomial equation.

Once we have time series data of the study variable for time points 1, 2, k the trend equations can be estimated using SPSS or any other statistical package. The SPSS can give the best-fit trend equation for a given set of data.

In many subject matter areas growth rates over time are required to be worked out, especially when yearly population, production, sale, profit, etc. are available. Normally growth rates can be linear or exponential (compound) based on the nature of the trend equation as linear or exponential.

For linear growth rate, the trend equation can be written as, $Y_t = Y_0 + bt$ where Y_t = the tth year value of the study variable, Y_0 is the starting value of the study variable and b is the rate of change (linear growth rate) of the study variable.

For exponential or compound growth rate the exponential growth model $Y = ab^{t}$ is assumed where Y and t are as given above; a = constant and b = 1 + r where r = exponential growth rate.

The growth rate calculation options are available under the statistical function in MS Excel.

17.4 Production Function

Production function in Economics is a physical relationship between output and input of any production system. In agriculture, the agricultural output of any crop (physical production of a production unit) is a function of many inputs (seed, fertilizer, labour, irrigation water, PP chemical, etc.) used in the production process. If Q stands for output and $X_1, X_2, \ldots, X_k$ stand for inputs, then the generally used linear and Cobb-Douglas production functions are of the form

$Q = a + b_1 X_1 + b_2 X_2 + \ldots + b_k X_k$ is the linear production function and $b_1, b_2 \ldots b_k$ are the marginal physical product of $X_1, X_2, \ldots, X_k$ with respect to output Q.

$Q = a x_1^{b1} X_2^{b2} \ldots X_k^{bk}$ is the Cobb-Douglas production function and b1, b2, . . ., bk are the elasticity of inputs $X_1, X_2, \ldots, X_k$ with respect to output Q.

Using the concept of output elasticity of input $= \frac{\text{Marginal Physical Product}}{\text{Average Physical Product}} = \frac{dQ/dX}{Q/X}$, the elasticity can be worked out from the estimated linear function and marginal products can be worked out from the Cobb-Douglas production function

17.5 Instability Analysis Based on Time Series Data

Measures of central tendency and measures of dispersion together reveal the pattern in a given set of data. The coefficient of variation ($CV = \frac{SD}{AM} \times 100$) is a scientific measure to assess internal consistency in cross-sectional set of data. Low CV implies high consistency in the data set.

But the CV value is not a good measure of consistency in a time series data set. We know that a time series dataset can have components like long-term trend (T), seasonal variation (S), cyclic variation (C) and irregular or random variation (R). These components can be in additive form or multiplicative form:

$Yt = T + S + C + R$ – Additive model

or

$$Yt = T \times S \times C \times R \quad - \text{ Multiplicative model.}$$

For most of the data especially economic data will have a positive trend for pushing factors and a negative trend for pulling factors. In other words, positive variables may have an increased trend and negative variables may have a negative trend. Hence the consistency in de-trended time series data can be assessed after eliminating the trend effect from original data. Using the estimated trend equation, we can calculate the trend values for each of the given data in the time series. If we assume an additive model, we have to subtract trend values from the original data and if assume a multiplicative model we have to divide the original data with trend values corresponding to each data. Let Y* denote the de-trended values of original data series Y, then we can calculate the CV of de-trended values as:

$$CV\left(Y^*\right) = \frac{SD\left(Y^*\right)}{AM\left(Y^*\right)} \times 100.$$

17.6 Linear Programming

Linear programming is a technique for getting optimum values-maximum or minimum-subject to certain constraints or conditions which limits the normal process to find such values. The term linear implies that the relationship involved in the problem must be linear (first-degree form of variables). The decision process for maximization of profit, minimization of cost, optimum allocation of resources and many other similar issues can find application of linear programming when such decisions are to be taken subject linear conditions. The basic assumptions of LP are:

- A goal for optimization as an objective function which can be expressed in a linear form of variables involved.
- Presence of structural conditions or constraints in the activities which again can be expressed in linear form of variables involved.
- In economic problems the prices are kept constant.
- Divisibility of activity levels in fractional or integer terms.
- Non-negativity of constraint variables.

Example: Assume that a production unit producing three products (X_1, X_2 and X_3). Let x_1, x_2 and x_3 be the number of units that are to be produced having prices p_1, p_2 and p_3. Suppose there is a capacity constraint (maximum possible at a time) for inputs in four uniform stages of the production process of each of the products. The problem can be written as in Table 17.3.

Table 17.3 Example of the Use of Linear Programming

Stages of Production Process	*Inputs Needed by the Product*			*Maximum Capacity of Units*
	X_1	X_2	X_3	
I	a_{11}	a_{12}	a_{13}	b_1
II	a_{21}	a_{22}	a_{23}	b_2
III	a_{31}	a_{32}	a_{33}	b_3
IV	a_{41}	a_{42}	a_{43}	b_4

The objective function (net revenue function that is to be maximized) is $R = x_1p_1 + x_2p_2 + x_3p_3$. The structural linear constraints are:

$$(a_{11} \times 1 + a_{12} \times 2 + a_{13} \times 3) \leq b_1$$

$$(a_{21} \times 1 + a_{22} \times 2 + a_{23} \times 3) \leq b_2$$

$$(a_{31} \times 1 + a_{32} \times 2 + a_{33} \times 3) \leq b_3$$

$$(a_{41} \times 1 + a_{42} \times 2 + a_{43} \times 3) \leq b_4.$$

The non-negativity constraints: $x_i \geq 0$.

Using slack variables, the inequality constraints are transformed into equality and the solution is worked out. A number of computer programming software like Visual Math, Gurobi Optimizer, General Algebraic Modeling System (GAMS), ILOG CPLEX Linear, Lingo, etc.

17.7 Input Response Function

Generally, input response functions for assessing the optimum level of use for inputs like fertilizer, irrigation, etc. assume great significance. According to the theory of marginal rate of substitution, if we go on increasing one input without making changes in other inputs the marginal return will go on increasing initially and then starts decreasing after a stage. Hence an inverted U-shaped curve for the production operates. In other words, an experiment with low to higher level of input (X) can be conducted and the production function will have a quadratic equation:

$$Y = a + bx + cx^2.$$

To find maximum of this production function, $\frac{dy}{dx} = 0$ and $\frac{d^2y}{dx^2}$ is −ve.

Hence $b + 2cx = 0$ or $x = -\frac{b}{2c}$ will give that level of input which maximizes production.

17.8 Decomposition Analysis

When we have data on two points of time for a time series variable like production say P_n and P_0. We may be interested to decompose the production change over time to area effect and yield effect.

We know that production $(P) = \text{Area } (A) \times \text{Yield } (Y)$.

$$P_n = A_n \cdot Y_n \text{ and } P_0 = A_0 \cdot Y_0$$

$$Pn - P0 = (An - A0)Y0 + (Yn - Y0)A0 + (An - A0)(Yn - Y0),$$

$$\text{that is, } 1 = \frac{(A_n - A_0)Y_0}{P_n - P_0} + \frac{(Y_n - Y_0)A_0}{P_n - P_0} + \frac{(A_n - A_0)(Y_n - Y_0)}{P_n - P_0}$$

that is,

$$1 \times 100 = \frac{(A_n - A_0)Y_0}{P_n - P_0} \times 100 + \frac{(Y_n - Y_0)A_0}{P_n - P_0} \times 100 + \frac{(A_n - A_0)(Y_n - Y_0)}{P_n - P_0} \times 100$$

Total change = Area effect + Yield effect + Area and Yield Interaction effect.

17.10 Total Factor Productivity

Total factor productivity (TFP) is a concept widely used in agriculture. Instead of factor productivity of individual inputs for any crop or agricultural production activity, the total factor productivity has a wider application for the selection of any production activity out of many alternatives possible as that activity with higher total factor productivity will be more rewarding.

Let Q is the quantity output per unit area of a crop. Let $x_1, x_2, \ldots, x_k$ are quantities of $X_1, X_2, \ldots, X_k$ inputs per unit area with prices $p_1, p_2, \ldots, p_k$, respectively. β is the TFP which is calculated:

$$\beta = \frac{Q}{p_1x_1 + p_2x_2 + \ldots + p_kx_k}$$

Using the above formula, the TFP for two alternative crops or for the same crop at two points of time can be compared. The crop with higher TFP will be more rewarding.

17.11 Diversification Index

Those who are engaged in agricultural research may come across problems of comparing agricultural diversification. The Simpson's diversity index (SDI) is a measure of diversity which takes into account the number of species present as well as the relative abundance of each species. As species richness and evenness increase the SDI will increase. The formula is set in such a way that the value of SDI will lie between zero and one. SDI = 0 means no diversity and SDI = 1 means very high diversity. SDI is calculated as:

$$\mathrm{SDI} = 1 - \frac{\Sigma n_i(n_i - 1)}{N(N-1)}$$

where n = total number of organisms of a particular species and N is the total number of organisms of all species.

As an example, if one is planning to calculate SDI for a region where different crop species of different crop commodities are grown as shown in Table 17.4.

Table 17.4 Example for Working Out Simpson's Diversity Index (SDI)

Crop Commodities	*Number of Species (n)*		*n(n−1)*	
	Area 1	*Area 2*	*Area 1*	*Area 2*
Cereals	6	3	30	6
Pulses	4	1	12	0
Oil crop	3	8	6	56
Spices	4	2	12	2
Vegetables	8	3	56	6
Fruits	2	0	2	0
Fodder	7	1	42	0
	N = 34	N = 18	160	70

$$\text{SDI for area } 1 = 1 - \frac{\sum n_i(n_i - 1)}{N(N-1)} = 1 - \frac{160}{34 \times 33} = 1 - (160/1122) = 1 - 0.14 = 0.86$$

SDI for area 2 = 1 – 70/18·17 = 1 – .23 = 0.77.

Here Area 1 is more diversified.

To sum up there are a large number of statistical univariate, bivariate and multivariate quantitative techniques applicable to data analysis in research. Depending upon the nature of the research outcome variable, aim and objectives, research hypothesis and availability of data the researcher can choose the most befitting quantitative techniques.

There can be many statistical quantitative techniques applicable to different research methods and designs based on the scope of study and availability of data. Some of the most basic statistical quantitative techniques that are applicable to different research methods and designs are summarized in Table 17.5.

Table 17.5 Research Methods, Research Designs and Some Basic Statistical/Analytical Techniques at a Glance Some Important points

Research Method	*Research Design*	*Some Possible Statistical Quantitative Techniques*
Experimental Method (applicable in health, agriculture, veterinary science, etc.)	True Experiment/ Randomized Control Trial	Mean, SD, CV, CI, test of significance (*t*- or F-test based on number of groups), correlation coefficient (r), regression analysis of outcome variable with independent variables.
	Post-Test Only	Mean, SD, CV, CI, test of significance (*t*-test), correlation coefficient (r), regression analysis of outcome variable with independent variables.
	Pre-Test Post-Test	Mean, SD, CV, CI, test of significance (paired *t*-test), correlation coefficient (r), regression analysis of outcome variable with independent variables.
	Solomon Four-Group	Mean, SD, CV, CI, test of significance (F-test), correlation coefficient (r), regression analysis of outcome variable with independent variables.
	Completely Randomized Design (CRD)	Mean, SD, CV, CI, ANOVA, CD, test of significance (F-test).
	Randomized Block Design (RBD)	Mean, SD, CV, CI, ANOVA, CD, test of significance (F-test).
	Latin Square Design (LSD)	Mean, SD, CV, CI, ANOVA, CD, test of significance (F-test).
	Factorial Design	Mean, SD, CV, CI, ANOVA, CD, test of significance (F-test).
	Quasi Experiment Design	Mean, SD, CV, CI, *t*-test.
	Non-Randomized Control Group Design	Mean, SD, CV, CI, test of significance (*t*-test).
	Time Series Design	Mean, SD, CV, CI, *t*-test between time points.
	Pre-Experiment Design	Mean, SD, CV, CI.
	One-Shot Post-Test Design	Mean, SD, CV, CI.
	One-Shot Pre-Test Post-Test Design	Mean, SD, CV, CI, paired *t*-test.

(*Continued*)

Table 17.5 (Continued)

Research Method	*Research Design*	*Some Possible Statistical Quantitative Techniques*
Observational/Non-Interventional Research Methods (clinical, para clinical and epidemiological)	Descriptive Case Reports	Description of the case in text form.
	Descriptive Case Series	Mean, SD, CV, CI.
	Descriptive Cross-Sectional	Mean, SD, CV, CI, correlation coefficient (r), regression analysis of outcome variable with independent variables.
	Descriptive Longitudinal	Mean, SD, CV, CI, trend analysis, correlation coefficient (r).
	Analytical Cross-Sectional	Mean, SD, CV, CI, test of significance (*t*-test), correlation and regression analysis.
	Observational Descriptive Studies	Mean, SD, CV, CI, test of significance (*t*-test).
	Analytical Case Control	Mean, SD, CV, CI, test of significance (*t*-test), odds ratio.
	Analytical Prospective Cohort	Mean, SD, CV, CI, test of significance (*t*-test), relative risk.
	Analytical Retrospective Cohort	Mean, SD, CV, CI, test of significance (*t*-test), odds ratio.
	Analytical Ambispective Cohort	Mean, SD, CV, CI, test of significance (*t*-test), odds ratio.
	Community Trials	Mean, SD, CV, CI, test of significance (*t*-test).
	Uncontrolled Natural Trials	Mean, SD, CV, CI.
	Before and After Comparison Trial/Impact study	Mean, SD, CV, CI, test of significance (paired *t*-test).
Clinical Observational Method (patient based)	Cross-Sectional Descriptive	Mean, SD, CV, CI, test of significance (*t*-test).
	Cross-Sectional Comparative	Mean, SD, CV, CI, test of significance (*t*-test).
	Cross-Sectional Correlation	Mean, SD, CV, CI, correlation, chi-square test.
	Cross-Sectional Exploratory	Mean, SD, CV, CI, correlation coefficient (r), regression analysis of outcome variable with independent variables.
Secondary Data based (all areas)	Spatial/Temporal Pattern of Parameters Under Study	Trend analysis, growth rates, instability analysis.
Qualitative Research Methods	Phenomenological Research	Descriptive textual explanations.
	Ethnographic Research	Descriptive textual explanations.
	Grounded Theory	Descriptive textual explanations.
	Action Research	Descriptive textual explanations.
	Narrative Research	Descriptive textual explanations.
Case Studies	'No Theory First' Type Case Study	Descriptive textual explanations.
	Multiple Case Studies Comparative Extreme Cases	Descriptive textual explanations.

Note: SD—standard deviation; CV—coefficient of variation; CI—confidence interval; CD—critical difference; ANOVA—analysis of variance, r—correlation coefficient

Some Important Points

- Kappa statistics is a measure of inter-observer agreement on the interpretation of situations like medical test results or similar events.
- Composite index is a pooling of different indices on objects so as to have a combined effect of such indices for inter-object comparison based on the pooled effect of indices.
- Trend is the pattern of the curve of any time series study variable.
- Production function in economics is a physical relationship between output and input of any production system.
- Linear programming is a technique for getting optimum values—maximum or minimum—subject to certain constraints or conditions which limits the normal process to find such values.
- Input response function is used for assessing the optimum level of use for inputs like fertilizer, irrigation, etc.
- Decomposition analysis is of varying types. The simplest is when we have data on two points of time for a time series variable like production say P_n and P_0. We may be interested to decompose the production change over time to area effect, yield effect and their interaction effect.
- Total factor productivity is the combined productivity of many factors of production.
- The Simpson's diversity index (SDI) is a measure of biodiversity which takes into account the number of spices present as well as the relative abundance of each species.

Suggested Readings

Berkman, H. W., and C. Gilson, *Consumer behavior: concepts and strategies*, 3rd ed., Kent, Boston, MA, 1986.

Box, G. E. P., G. M. Jenkins, and G. C. Reinsel, *Time series analysis: forecasting and control*, 3rd ed., Prentice Hall, Englewood Cliffs, NJ, 1994.

Brockwell, P. J., and R. A. Davis, *Time series: theory and methods*, 2nd ed., Springer-Verlag, 1991.

Cryer, J. D., *Time series analysis*, Duxbury Press, Boston, MA, 1986.

Fuller, W. A., *Introduction to statistical time series*, John Wiley & Sons, New York, 1976.

Gottman, J. M., *Time-series analysis: a comprehensive introduction for social scientists*, Cambridge University Press, Cambridge, 1981.

Harvey, A. C., *The econometric analysis of time series*, Philip Allan, Oxford, 1981.

Harvey, A. C., *Forecasting, structural time series models and the Kalman filter*, Cambridge University Press, Cambridge, 1989.

Koutsoyiannis, A., *Theory of econometrics*, Macmillan, London, 1976.

Kraemer, H. C., 'Kappa coefficient', in *Encyclopedia of statistical sciences*, edited by S. Kotz and N. L. Johnson (eds.), John Wiley & Sons, New York, 1982.

Loudon, D. L., and A. J. Della-Bitta, *Consumer behavior concepts and application*, 4th ed., McGraw-Hill, New York, 1993.

Pena, D., G. C. Tiao, and R. S. Tsay (eds.), *A course in time series analysis*, John Wiley & Sons, New York, 2001.

Solomon, M. R., *Consumer behavior: buying, having and being*, 2nd ed., Allyn & Bacon, Boston, MA, 1994.

18 Computer Application in Research

18.1 Introduction

Computer application has become inevitable in various activities of research. Starting from research topic identification to research publication, ICT application plays a crucial role and eases the research activities at various stages as shown in Figure 18.1.

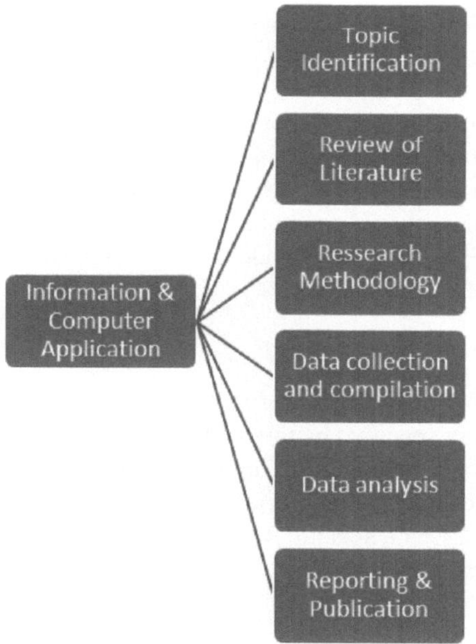

Figure 18.1 Computer Application in Research Activities

Topic Identification: The website visits of related research institutes and internet searches for important research topics in the area of interest of the researcher will help the researcher to set the research platform for his/her proposed research work. After having a panel of possible research topics, the researcher can shortlist topics using the FINER (feasibility, interest, novelty, ethics and relevance) criteria. After discussions with theresearch guide/research board of the researcher, he/she can reach a decision on the specific topic of study.

DOI: 10.4324/9781003527183-18

Review of Literature: The ICT application makes it possible for the researcher to have a wide range of online reviews of literature of research journals, digitally published books and other materials. In fact, most of the researchers initiate review of literature even before topic finalization as it helps them to identify novel research topics. A comprehensive literature review will help the researcher right from topic identification to report finalization and publication of results. A systematic review will always enhance the quality and relevance of research.

Research Methodology: While reviewing past research work related to a specific area/topic the researcher develops insights for the methodology of his/her identified topic of study. With required modifications/improvements the research methodology for the identified topic can be streamlined by the researcher.

Data Collection and Compilation: In most research studies the data collection is done manually or with the help of specially designed machines depending upon the research methods as experimental, survey-based, observational or other methods including epidemiological studies. Remarkably, in clinical research, the data of ICU patients are collected from computer-based equipment/devices. The Google forms are being used to collect opinion/survey-based data from educated and computer-literate respondents. Online surveys are gaining dominance these days in market research and other areas.

Data Analysis: Google Form-based collected information can be easily transferred to Excel sheets. The manually collected data can also be entered on Excel sheets either for preliminary analysis using available options in Excel or for software-based analysis using SPSS/SAS/R-programme, etc. as most of these programmes allow import/export options from Excel worksheets for further data analysis.

Reporting and Publications: MS Word has a wide range of options to prepare reports using text matter, graphs, tables, charts, diagrams and other editing including plagiarism-checking options of the manuscript. As there is a wide range of online journals, publication has become faster and easier with ICT applications.

18.2 MS Excel Worksheet

MS Excel is a multi-purpose user-friendly programme for researchers. Every unit of data occupies a cell having its identity specified by the row and column number. It has 1048576 rows and 16484 columns. The column width can go up to 255 characters and row height can have 409 points. The dependent and independent variables are generally put in columns in the dataset. Each unit of study can be given a specific row with the unit code preferably. Basic knowledge of MS Excel has many advantages.

- The worksheet settings can be made by the researcher once the synopsis is ready with the data collection format.
- The data entry can be simultaneously done with data collection and the problem of data inconsistencies and missing data can be minimized.
- The manual validation of data in terms of permissible data range (outlier problems), missing data, etc. can be done during data entry and by visual screening of data.
- Basic statistical data analysis can be done using inbuilt options of Excel under < Formulas> < More Functions> <Statistical> or by putting = in a cell and selection of appropriate function.
- It has facility to set and edit formulas based on cell identity and get results instantly for the dataset or any part of the marked dataset.

The statistical functions may vary with the version of MS Office but some of the inbuilt options for getting statistical results from the dataset of Excel are given in Table 18.1 and 18.2.

Table 18.1 Some Functions Available in Excel

	Is Function	*Conditional*	*Mathematical*	*Find and Search*	*Lookup*	*Reference*	*Date and Time*	*Misc.*	*Rank*	*Logical*
1	*ISBLANK*	*AVERAGEIF*	*COUNT*	*FIND*	*MATCH*	*ADDRESS*	*DATE*	*AREAS*	*RANK*	*AND*
2	*ISERR*	*AVERAGEIFS*	*COUNTA*	*SEARCH*	*LOOKUP*	*CHOOSE*	*DATEVALUE*	*CHAR*	*RANK.AVG*	*OR*
3	*ISERROR*	*SUMIF*	*COUNTBLANK*	*SUBSTITUTE*	*HLOOKUP*	*INDEX*	*TIME*	*CODE*	*RANK.EQ*	*XOR*
4	*ISEVEN*	*SUMIFS*	*AVERAGE*	*REPLACE*	*VLOOKUP*	*INDIRECT*	*TIMEVALUE*	*CLEAN*		*NOT*
5	*ISODD*	*COUNTIF*	*AVERAGEA*			*OFFSET*	*NOW*	*TRIM*		
6	*ISFORMULA*	*COUNTIFS*	*MEDIAN*				*TODAY*	*LEN*		
7	*ISLOGICAL*	*IF*	*MOD*				*YEAR*	*COLUMN*		
8	*ISNA*	*IFERROR*	*SUM*				*MONTH*	*ROW*		
9	*ISNUMBER*	*IFNA*	*SUBTOTAL*				*DAY*	*EXACT*		
10	*ISREF*		*SUMSQ*				*HOUR*	*FORMULATEXT*		
11	*ISTEXT*		*SUMPRODUCT*				*MINUTE*	*LEFT*		
12	*ISNONTEXT*		*SQRT*				*SECOND*	*RIGHT*		
13			*POWER*				*WEEKDAY*	*MID*		
14			*EVEN*				*DAYS*	*LOWER*		
15			*ODD*				*NETWORKDAYS*	*PROPER*		
16			*INT*				*WORKDAYS*	*UPPER*		
17			*LARGE*				*WORKDAY*	*REPT*		
18			*SMALL*					*SHEET*		
19			*MAX*					*SHEETS*		
20			*MAXA*					*TRANSPOSE*		
21			*MIN*					*TYPE*		
22			*MINA*					*VALUE*		
23			*RAND*							
24			*RANDBETWEEN*							

Table 18.2 Some Important Statistical Functions in MS Excel

Functions	*Corresponding Statistical Values*
AVEDEV	Average of absolute deviations
AVERAGE	Average of values
BINODIST	Individual binomial distribution probability
CHIDIST	Probability of chi-square distribution
CHITEST	Value of chi-square distribution for the statistic as per DF
CONFIDENCE	Confidence interval for population mean
CORREL	Correlation coefficient for two sets of data
COUNTIF	Count the number of cells within a given range
COVAR	Covariance of two sets of data
DEVSX	Sum of squares of deviations of data from sample mean
FISHER	Fisher's transformation of data
GEOMEAN	Geometric mean of +ve numerical data
GROWTH	Exponential growth rate of a set of data
HARMEAN	Harmonic mean of a set of data
INTERCEPT	Intercept value of best fit linear regression for X,Y data
KURT	Kurtosis for a set of data
LARGE	K^{th} largest value in a set of data
LINEST	Linear trend for a given data set
LOGEST	Exponential trend for a set of data
MAX	Largest value in a set of data
MEDIAN	Median for a set of data
MIN	Smallest value in a data set
MODE	Mode of a set of data
NORMDIST	Normal cumulative distribution for specified mean and SD
NORMSDIST	Normal cumulative distribution for SND
PERCENTILE	K^{th} percentile of a given set of data
POISSON	Poisson distribution value
PROB	Probability of values in a given range
QUARTILES	Quartiles of the data
RANK	Rank of a data in a given set
SKEW	Skewness of a distribution
SLOPE	Slope of linear regression for a given set of data
SMALL	K^{th} smallest value in a given set of data
STANDARDIZE	SND values for a given set of data
STDEV	Standard deviation based on a sample
STDEVA	Estimated standard deviation based on a sample
STDEVP	Standard deviation of a population data
STEYX	Standard error of the predicted Y values
TREND	Linear trend of a set of time series data
TTEST	Probability associated with standard *t*-test
VAR	Estimated variance based on sample data
VARP	Variance of population data
ZTEST or Z.TEST	P value of one tail Z-test

One can get any sort of help from exceldemy.com or search on a search engine like google.com. Help is also available in Excel itself. After typing '=' and some of the alphabets of the required function, a list of related functions appears and one can select the desired function for use.

One can easily look for help with Excel on any of the search engines.

Availability of functions may vary with the version of the MS Office. Using a function in Excel is very easy. You may select a function either by typing '=' and some alphabet related to the function, a list of functions will appear, select a function (Figure 18.2) or click on '*fx*', a box will popup, select the type of function (Figure 18.3) either by typing the needed function in the shaded area of Figure 18.4 or selecting a category of function then choosing one from the list of functions as can be seen in Figure 18.4.

Enter the required information in the brackets () as desired by the function in the order, for example, for calculating average, size of the array, is required =AVERAGE(A1:A10). It will return the average of A column from rows one to ten. You may enter a block of any number of columns and rows =AVERAGE(A1:D10) or different blocks =AVERAGE(A1:B10,E5:G15,K7:N21).

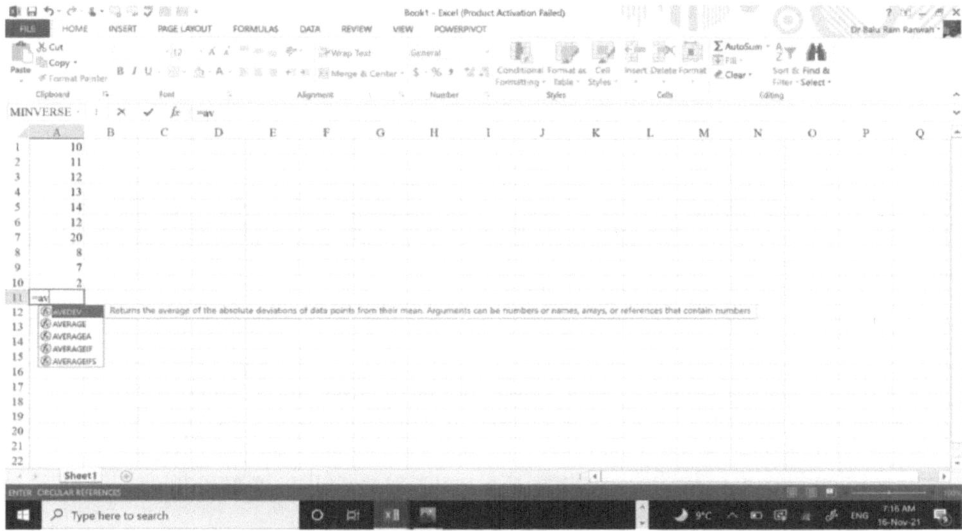

Figure 18.2 Selection of Functions Using '='

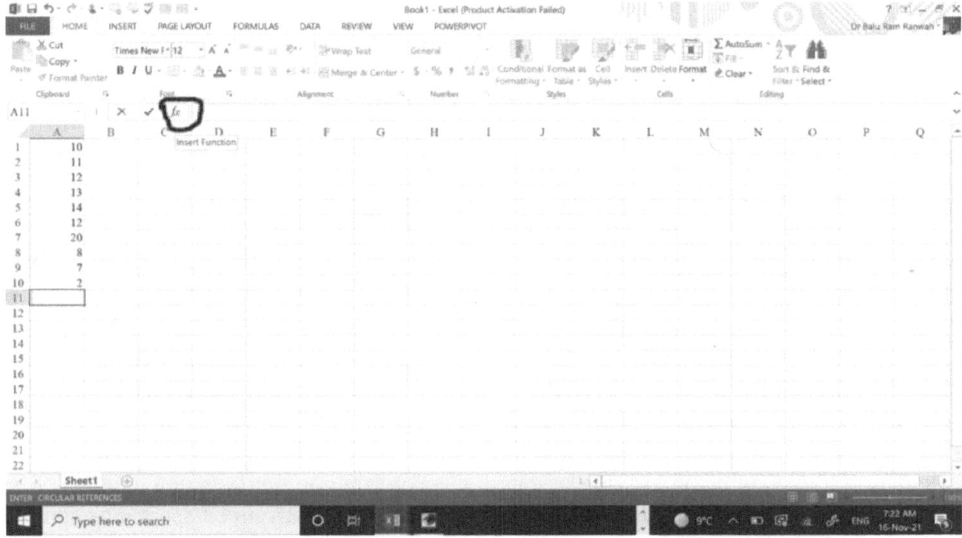

Figure 18.3 Selection of Functions Using '*fx*' in Excel

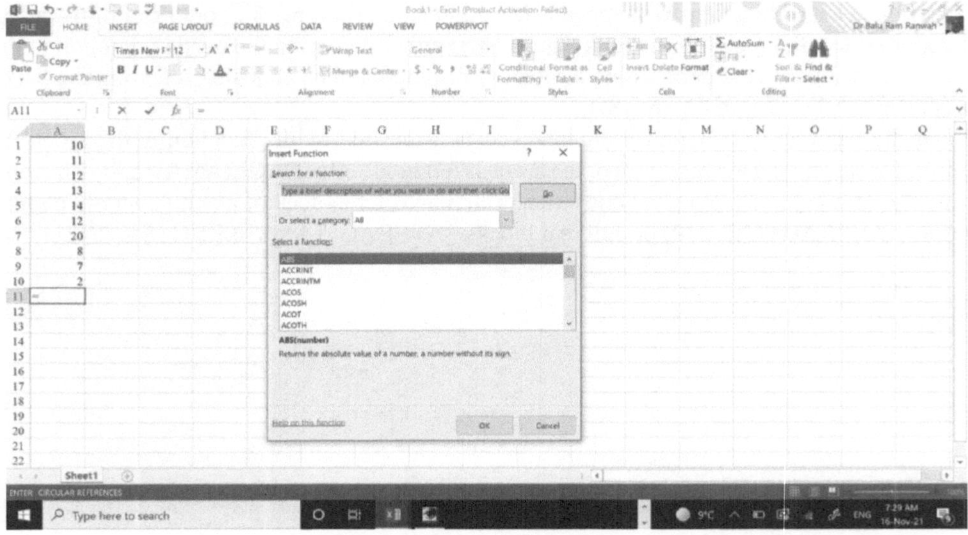

Figure 18.4 Pop-Up Box for Selection of Functions in Excel

In this way, we can perform most of the calculations by selecting the appropriate function in the appropriate cell and value. For the same size of data same function can be copied and the function can be repeated. When we are having a number of data sets of different sizes calculation becomes cumbersome. In this situation, it is better to use other programmes like SPSS, R, SAS, etc.

Inbuilt options are available in Excel for doing basic statistical analysis. However, for advanced statistical analysis, one has to use appropriate computing software with care and caution as 'garbage in is garbage out' for any partial knowledge application of advanced data analysis software. A large number of software are available for statistical analysis. Most of them are customized for specific fields of specialization. Some of the generalized popular programmes are SPSS, R-programme and SAS. The SPSS is most popular because of its user-friendly nature.

18.3 SPSS in Data Analysis

This is most popular software among researchers of social sciences and provides solution for most of the problems. To facilitate the researchers, to date 30 versions have been released by the company. It has four types of files: (1) Data file, (2) syntax file, (3) output file and (4) script file. The data file has the datasets. Syntax file has the necessary function with specifications and conditions to be used for the calculation in a specific order. Selected functions can be applied to the selected set of data. All results are stored in the output file. The output file is a passive file which can be saved or exported in different formats. Script files consist of codes in languages like Basic, Python, etc. Basic is already integrated within SPSS whereas for Python, one is required to install Python and SPSS-Python intercession plug-in. Scripting helps in automating the calculation like opening and saving of data files, exporting output files in a number of formats and customizing output in the viewer. Scripting is only for advanced users. For common researchers, only data files and output files are used. Some of the researchers also use the syntax.

Different functions are grouped and listed in menu bar ribbons like File, Edit, View, Data, Transform, Analysis, Graphs, Utilities, Add-ons, Windows and Help (Figure 18.5).

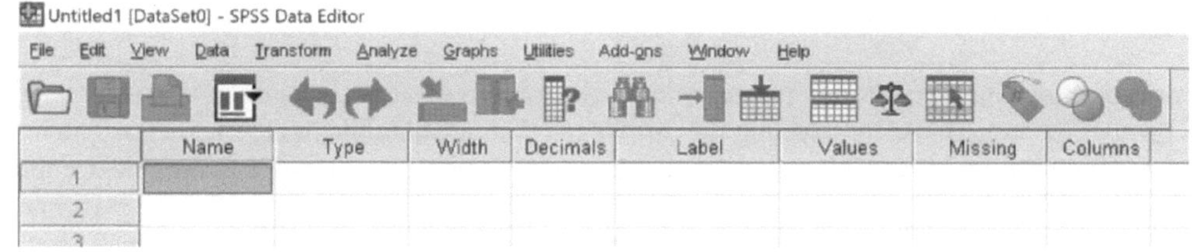

Figure 18.5 Menu Bar of SPSS

When we click on any menu or ribbon, a list of functions is available in each menu (Figure 18.6). Details of each menu are:

File: All file-related functions are available in this menu such as open, save, save as etc.
Edit: To edit any file, all functions like cut, copy, paste, etc. are available in this menu.
View: What is to be viewed while using the data file is to be selected from this menu.
Data: All data-related functions are available in this menu.
Transform: The functions related to generate new column or replace the column with the help of existing data of column(s) are available in this menu.
Analyze: This is the most important menu as all functions related to statistical analysis of data are available in this menu.
Graphs: All options to develop different types of graphs with the help of existing data functions are available in this menu.
Utilities: Other related functions are available in this menu.
Add-Ons: Applications, services, statistical guides, etc. are available in this menu.
Window: The files opening and minimize or maximize functions are available in this menu.
Help: All sorts of help are available in this menu.

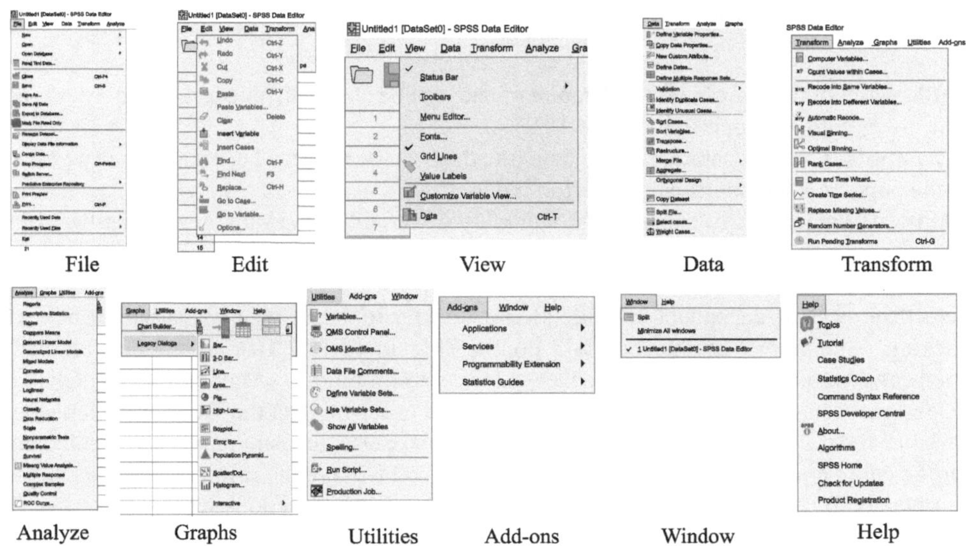

Figure 18.6 Different Options Under Each Menu on the Menu Bar of SPSS

To use this programme we have to follow these steps:

Preparation of data file: We may create the data file in SPSS itself or may import the file prepared in any other programme by selecting the format and then opening file in the file menu. One may have data for different variables recorded on a number of samples or plots. Each variable occupies one column. Values recorded on the sample are placed in rows in respective cells of that variable. Number of file management tools are available in the software like splitting the data, merging the data, calculating new variables with the help of existing variables, changing the nature of data (nominal, ordinal and scale). The data file has two views: (a) Data view and (b) variable view.

Data view: In this, all data of different variables are visible. We can edit the data in this view (Figure 18.7).

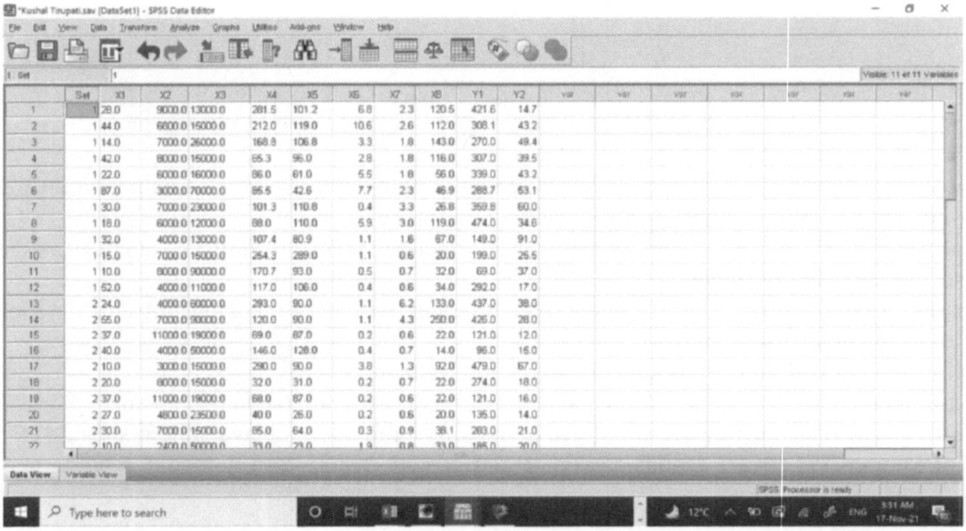

Figure 18.7 Data View of SPSS File

Variable view: In this view name and nature of data can be changed. One can create a new variable along with its properties (Figure 18.8).

Data Analysis: After preparation of the data file, data can be analyzed using the analyze menu. Some popular calculations can be performed as:

Summary of Variables: To calculate the summary of variables like sum, mean, number, minimum, maximum, standard deviation, skewness, kurtosis and variance of variables. Select <Analyze>, <Reports> and <Report summaries in row> a box will pop up (Figure 18.9). Select the desired dependent variable(s) and transfer to <data column variable> box. On clicking <Summary> another box will pop up (Figure 18.10). Tick on the desired calculated values required for the variable like <Mean>, <Minimum>, <Maximum>, etc. Click on <Continue>. If summaries are required for different groups, select the grouping independent variable and transfer in <Break Column Variable> box else, the summary will be calculated for the entire column of variable. Click on <OK>. You will get the result. The same can also be obtained by <Analyze>, <Reports>, <Case Summaries> functions where the rest of the procedure is the same except instead of ticking a type of summary required click on <Statistics> select desired parameter and transfer in <Cell Statistics> box (Figure 18.11), click on <Continue> this will disappear, click on <OK>. Resultant statistics will be in the output file.

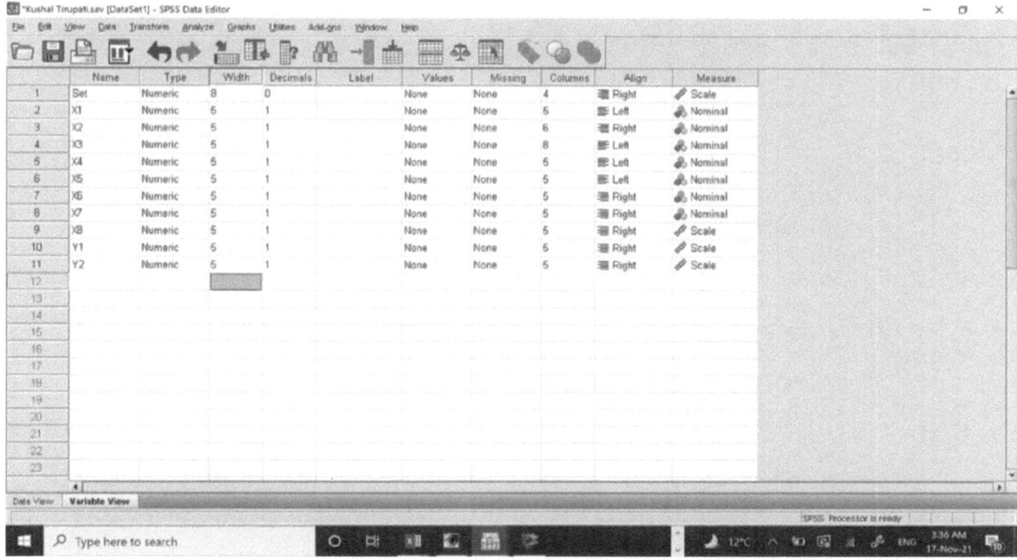

Figure 18.8 Variable View of SPSS File

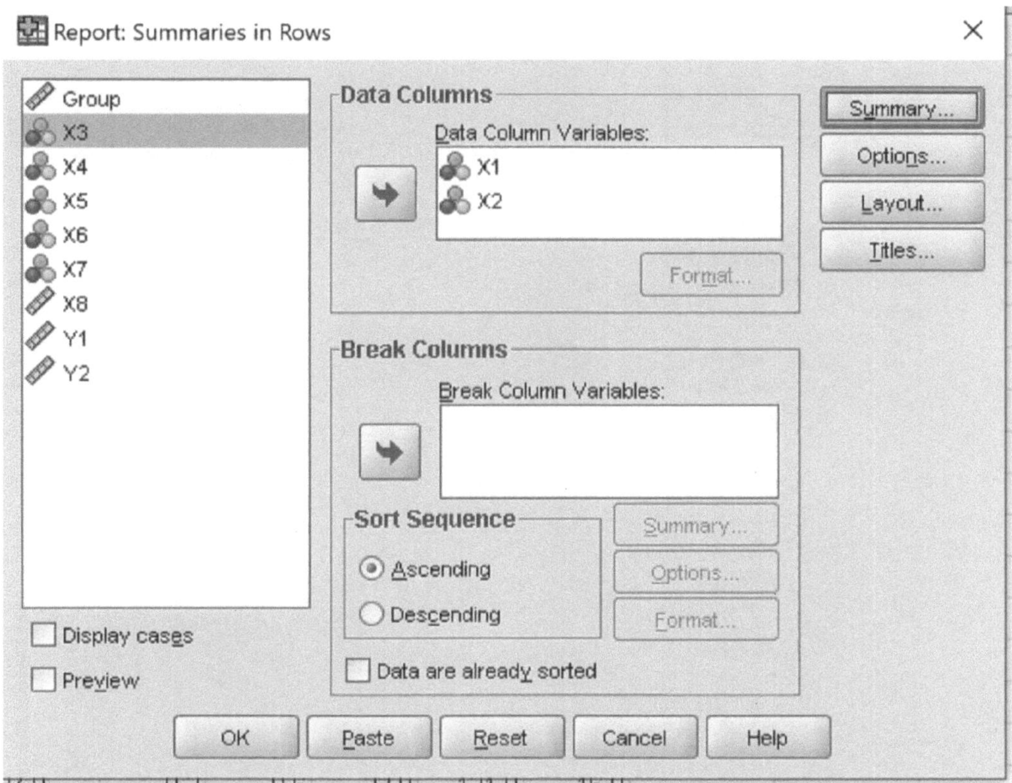

Figure 18.9 Report Summaries in Row Pop-Up Box in SPSS

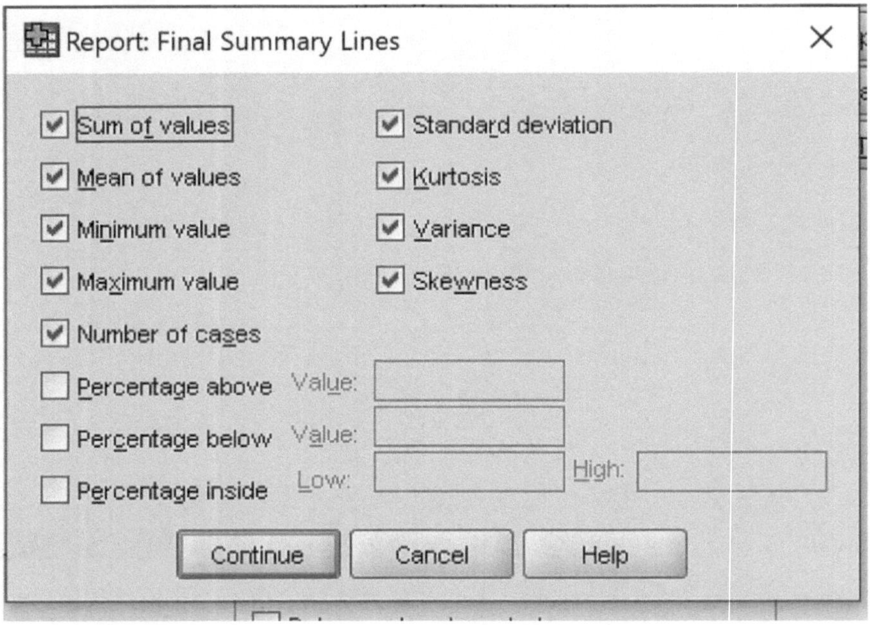

Report: Final Summary Lines
Sum of values
Mean of values
Minimum value
Maximum value
Number of cases
Percentage above
Percentage below
Percentage inside
Standard deviation
Kurtosis
Variance
Skewness
Value:
Value:
Low:
High:
Continue
Cancel
Help

Figure 18.10 Summary Pop-Up Box in SPSS

Summarize Cases
Summary Report: Statistics
Statistics:
Median
Grouped Median
Std. Error of Mean
Sum
Minimum
Maximum
Range
First
Last
Standard Deviation
Variance
Kurtosis
Std. Error of Kurtosis
Cell Statistics:
Number of Cases
Mean
X3
X4
X5
X6
X7
X8
Y1
Y2
Statistics...
Options...
Continue
Cancel
Help

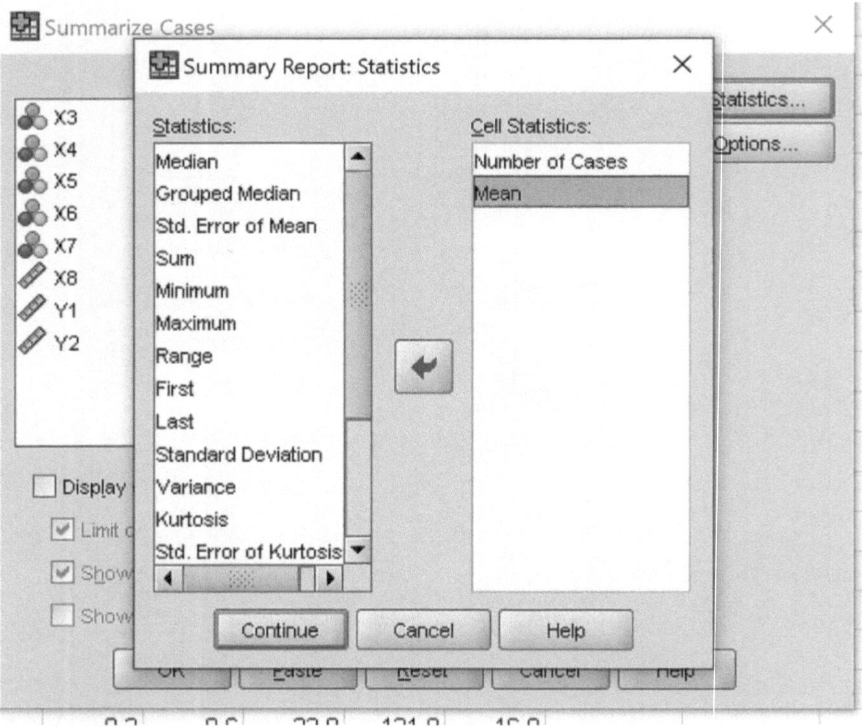

Figure 18.11 Summary Report Statistics Pop-Up Box in SPSS

Crosstabulation: Click on <Analyze>, <Descriptive Statistics>, <Cross Tabs> a box will popup select grouping variable and transfer in <Column(s)>, select another variable and transfer in <Row(s)>, the transfer can be vice versa, click on <Statistics> select desired statistics, <Continue>, this box will disappear, click on <Cells> to obtained cell value to be obtained like Count-Observed and/or Expected, Percentage- Row, Column and/or Total, etc. click <Continue> (Fig. 18.12) the box will disappear, click on <OK> result will be in output file.

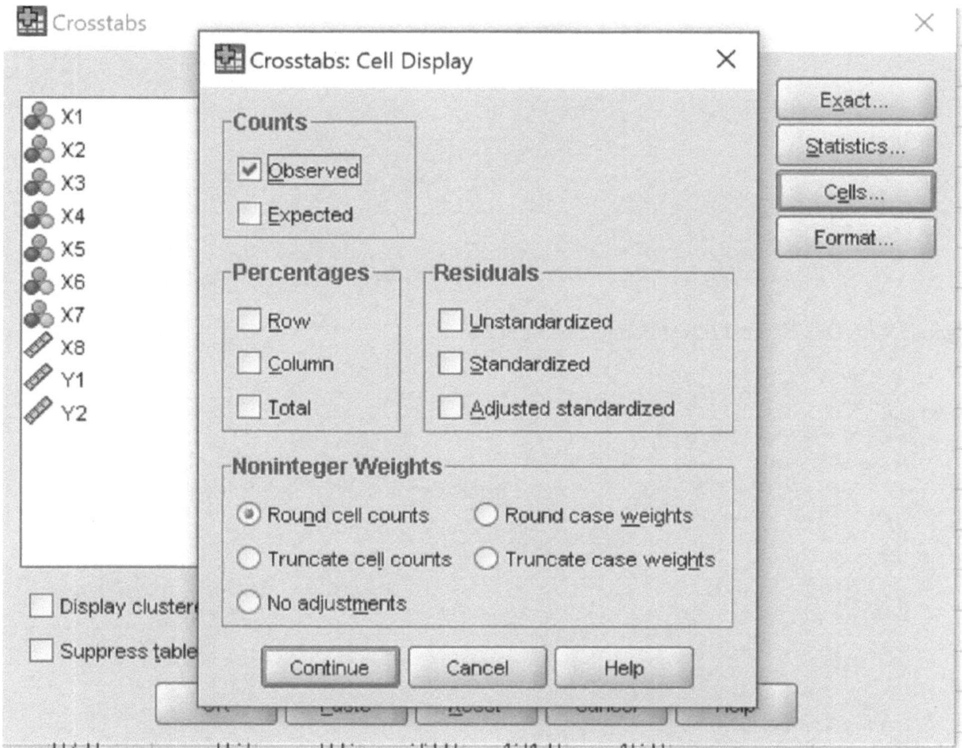

Figure 18.12 Cross Tabs Cell Display Pop-Up Box in SPSS

Compare means: Means can be compared by one sample *t*-test, independent sample *t*-test, paired sample *t*-test or one-way ANOVA depending on the situation. Click on <Analyze>, <Compare Means>, select desired statistics like <One-Sample *t*-test> a box will pop up, select the test variable like plant height and transfer in <Test Variable(s)> box, put test value if available else it will compare with zero, click on <OK>, results will be there in the output file (Figure 18.13). Similarly, other statistics can also be obtained by selecting the appropriate function. Up to one-way ANOVA (simple CRD) is possible in <Compare Means> menu for other complicated mean comparisons <General Linear Model> or <Generalized Linear Models> menu may be used.

Correlation analysis: To calculate the correlation between a set of characters click on <Analyze>, <Correlate> and <Bivariate> a box will popup, select the set of characters and transfer in the variable box, select the type of correlation to be applied like Pearson/Kendall's Tau-b/Spearman then select the test of significance (one-tailed or two-tailed) then select <Flag significant correlations> then click <OK> (Figure 18.14), results will be in output file.

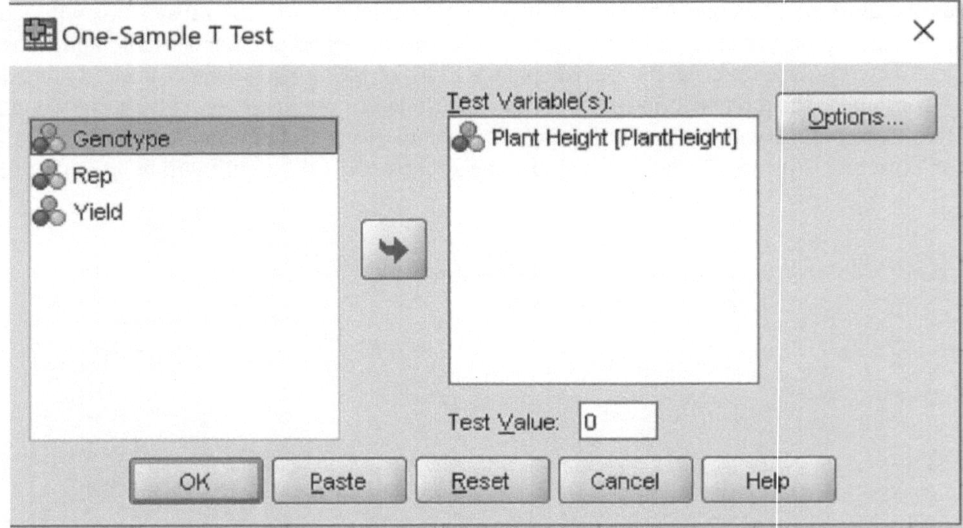

Figure 18.13 One-Sample *t*-Test Popup Box in SPSS

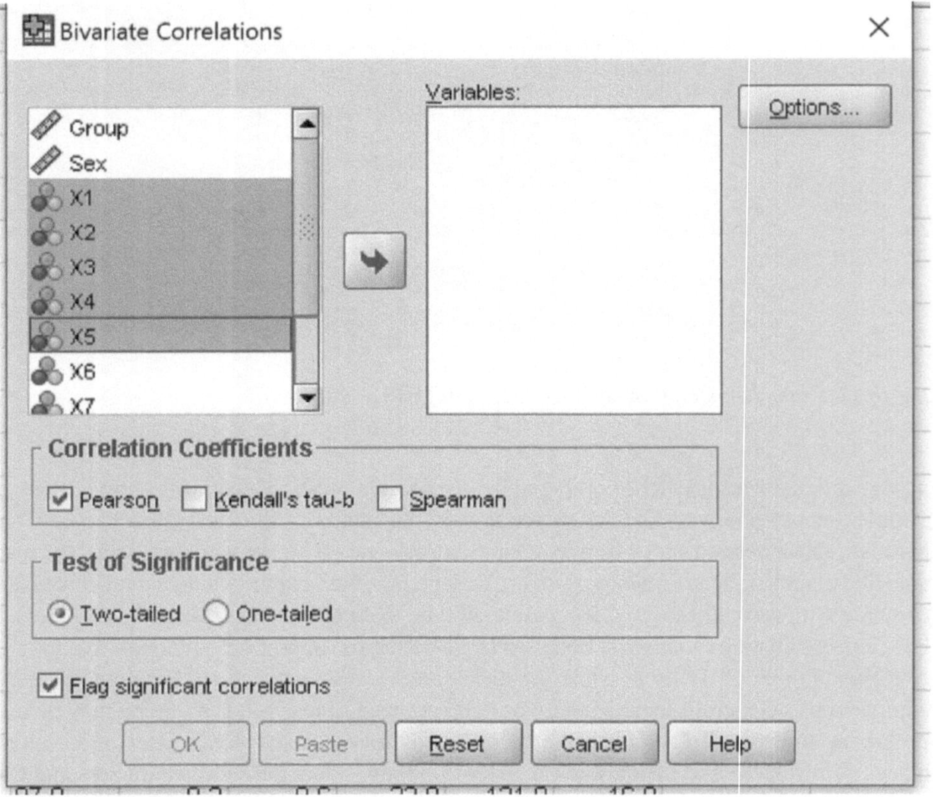

Figure 18.14 Pop-Up Box for Bivariate Correlation Analysis in SPSS

Similarly, other statistics can also be obtained by selecting the appropriate function.

18.4 R Programme

R is an interactive programming language has most of the statistical and graphical functions. It has more flexibility of data input and output. Output format can be changed as per requirement but is for advanced users having the required knowledge of programme writing. It is a command-based programme. Screen shot of R is given in Figure 18.15.

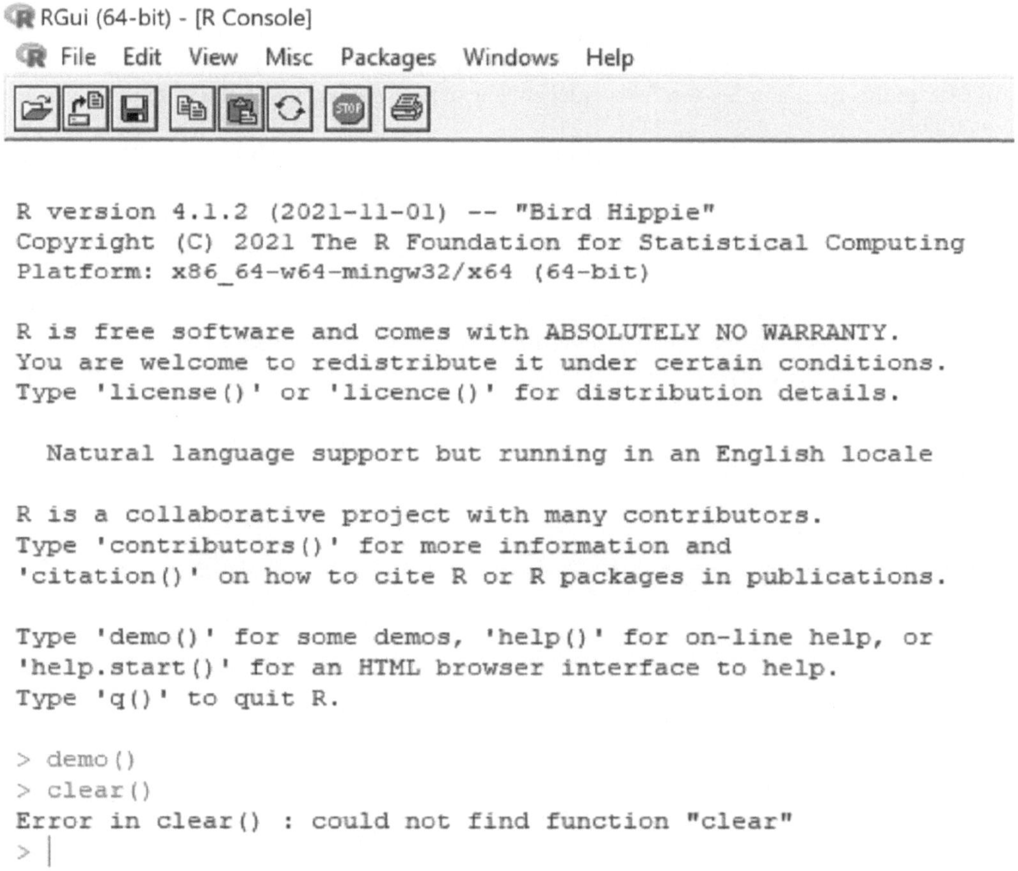

RGui (64-bit) - [R Console]

File Edit View Misc Packages Windows Help

```
R version 4.1.2 (2021-11-01) -- "Bird Hippie"
Copyright (C) 2021 The R Foundation for Statistical Computing
Platform: x86_64-w64-mingw32/x64 (64-bit)

R is free software and comes with ABSOLUTELY NO WARRANTY.
You are welcome to redistribute it under certain conditions.
Type 'license()' or 'licence()' for distribution details.

  Natural language support but running in an English locale

R is a collaborative project with many contributors.
Type 'contributors()' for more information and
'citation()' on how to cite R or R packages in publications.

Type 'demo()' for some demos, 'help()' for on-line help, or
'help.start()' for an HTML browser interface to help.
Type 'q()' to quit R.

> demo()
> clear()
Error in clear() : could not find function "clear"
> |
```

Figure 18.15 Command Window of R

18.5 SAS Programme

It is a very powerful programme having a number of modules for researchers of different disciplines. It is a menu as well as a command-based programme. It has an active output file whose results can be reused by the programme to calculate new values. But, to use it, one needs very high skill due to its use by advanced users only. Though, it has functions for most calculations, it is not very popular among common researchers.

18.6 JAMOVI Programme

It is open-source software freely available on www.jamovi.org website under the license AGPL3. It is available for all operating systems like Windows, macOS, Linux and Chromebook. It can be installed on operating systems from Windows Vista (64-bit) onward. It is menu-based software similar to SPSS. Different programmers developing the functions and uploading on the site are available for all under JAMOVI library. R programmes can also be run in JAMOVI. All help, functions, codes, etc. are available on this website under the resource menu. Details help on exploration, *t*-tests, ANOVA, regression, frequencies and factors are available. Programmes available under each heading are:

- Exploration: It is provided under the heading *Descriptive* where information about data like count, mean, minimum, maximum, etc. are available.
- *t*-Test: In this one sample *t*-test, two independent sample *t*-tests and paired sample *t*-tests are available.
- ANOVA: In this one-way ANOVA, ANOVA, repeated measures ANOVA, ANCOVA, MANCOVA, one-way ANOVA (non-parametric) and repeated measures ANOVA (non-parametric) are available.
- Regression: Correlation matrix, partial correlation, linear regression, binomial logistic regression, multinomial logistic regression and ordinal logistic regression are available.
- Frequencies: Proportion test (two outcomes), proportion test (N outcome), contingency table, paired samples contingency table and log-linear regression are available.
- Factor: Reliability analysis, principal component analysis, exploratory factor analysis and confirmatory factor analysis.

Window of JAMOVI is given in Figure 18.16.

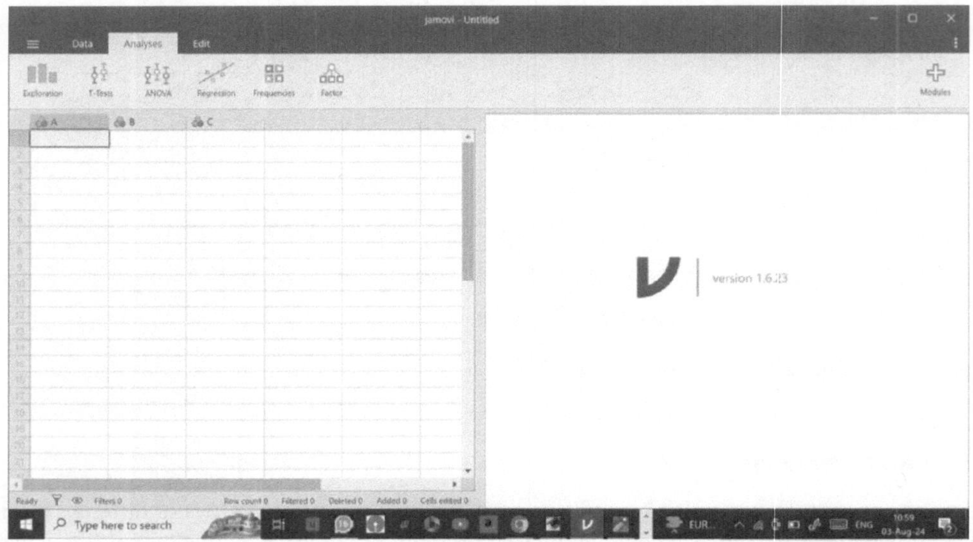

Figure 18.16 JAMOVI Window

18.7 Application of Artificial Intelligence in Research

Artificial Intelligence (AI) is intelligence exhibited by machines and softwares. AI is coming up as a general purpose technology that has a large number of applications in all spheres of life including language translation, image recognition, decision-making, e-commerce, genetic manipulations and many others. It also plays a crucial role in education sector. In research, in different areas of study, it is expected that AI will make its intervention both in quality and quantity of research. The application of AI in research can have its role right from identification of most relevant topic to publication of research results in journals of high repute. In the near future, intelligent machines may replace or enhance human capabilities in many areas.

Some Important Points

- Starting from topic identification for research to publication, ICT application eases the research activities at various stages.
- MS Excel is a multi-purpose user-friendly programme for researchers. Every unit of data occupies a cell having its identity specified by the row and column number.
- Basic statistical data analysis can be done using inbuilt options of Excel under <Formulas> < More Functions> <Statistical> or by putting = in a cell and selecting the appropriate function.
- SPSS is the most popular software among researchers in various fields which provides solutions for most of the problems. To facilitate the researchers, to date, 28 versions have been released by the company. It has four types of files: (1) Data file, (2) syntax file, (3) output file and (4) script file. Different functions are grouped and listed in the menu bar ribbon like File, Edit, View, Data, Transform, Analysis, Graphs, Utilities, Add-ons, Windows and Help.
- R is an interactive programming language having most of the statistical and graphical functions. It has more flexibility of data input and output.
- SAS is a very powerful programme having a number of modules for researchers of different disciplines. It is a menu-based as well as a command-based programme.
- JAMOVI programme is an open-source software freely available on www.jamovi.org website under the license AGPL3. It is available for all operating systems like Windows, macOS, Linux and Chromebook.
- The application of AI in research can have its role right from identification of most relevant topic to publication of research results in journals of high repute.

Suggested Readings

Burns, P. R., 'Multiple comparison methods in MANOVA', in *Proceedings of the 7th SPSS users and co-ordinators conference*, Northwestern University, Evanston, IL, 1984.

Dixon, W. J., *BMD biomedical computer programs*, University of California Press, Los Angeles, CA, 1973.

Horton, N. J., and S. R. Lipsitz, 'Review of software to fit generalized estimating equation regression models', *The American Statistician* 53: 160–169, 1999.

Knuth, D. E., *The art of computer programming, vol. 3: sorting and searching*, Addison-Wesley, Reading, MA, 1973.

Lee, E., and M. Desu, 'A computer program for comparing k samples with right censored data', *Computer Programs in Biomedicine* 2: 315–321, 1972.

Watts, D. L., 'Correction: computer selection of size-biased samples', *The American Statistician* 45(2): 172–172, 1991.

19 Writing Research Synopsis, Thesis and Papers

19.1 Introduction

Documentation is an integral part of research. Right from preparation of the synopsis/research protocol to periodical reports ending with the final report/thesis and finally publication of research papers requires the skill for scientific documentation. Precision and clarity are two factors which enhance the utility of scientific writing. Research-related writings require a scientific mode of thinking and application of scientific methods. Hence critical and creative mind is a necessary requirement for any researcher to make his scientific writing more cohesive, precise and concise.

19.2 Writing Research Protocol/Synopsis/Proposal

A research protocol or a research synopsis or a research proposal is a document submitted by the researcher to an authority or institution for formal approval, ethical clearance, formal registration in universities for the award of a research degree, financial assistance from organizations/funding agencies, etc. Preparation of a winning research proposal is an art as well as a science. Synopsis is the outline or road map of a planned project submitted for approval from competent authorities. It gives clarity and direction of proposed research work to the researcher and a panoramic view of the research for quick understanding by the peer reviewers.

Thus, a protocol or a synopsis forms a crucial part of a research project or a thesis. Many universities have made it mandatory for postgraduate degree students to prepare a synopsis as a part of their postgraduate training. Good knowledge about how a protocol or a synopsis is written is imperative to all researchers involved in different areas of research.

Literally, protocol comes from the Greek word 'protokollo' meaning a format procedure for carrying out scientific research. Similarly, synopsis means a brief summary plan of something. Frequently, both the terms are used as synonyms, but the term 'synopsis' is more common.

A synopsis should be constructed in such a manner that it facilitates the reviewer/expert to understand the research project at a glance. The researcher's clarity of perception about the research goals/objectives and the strategies to achieve the goal is reflected in the synopsis. It should be brief but precise. A well-conceived synopsis makes the task easier for the researcher. The synopsis is generally structured under headings like title, statement of the problem and hypothesis, aims and objectives, review of literature, research methodology and references. A research proposal for external support/funding will have all these items along with information revealing the researcher's institutional affiliation and his/her biodata depicting his/her competencies, etc. A brief note on each item of a research synopsis is discussed:

DOI: 10.4324/9781003527183-19

19.2.1 Title/Topic of Research

The title of the research project should be brief but self-explanatory and depict the focal theme addressed for research in ten to 12 keywords. It can also be in two parts. The first part consists of keywords depicting the focal theme of the study and the second part a few qualifying words to specify the special features of the study, for example, 'Impact of Organic Agriculture on Income and Employment in Rajasthan—A Rural Perspective'.

19.2.2 Importance/Statement of Problem of Study

It can also be titled 'Introduction' or 'Problem Formulation' or 'Importance of Study'. The researcher will have to justify his/her choice to work on the particular topic with the definitions of keywords in the title, facts and figures related to the topic and the stipulated objectives that flow from the description concisely in a few pages. This section may include the following points:

- The relevance and contemporary importance of identified research problem/study should be highlighted.
- The relevance of the topic at global, national, state and local levels may be highlighted. A brief account of its utility at the local or national level must be discussed.
- The present status of work on the problem, emerging issues and the need for taking up the study to be justified.
- Recently carried out studies on similar lines and how far the present study differs from those may be highlighted.
- The research hypothesis should be stated as precise and concise statements on the expected outcome of the study reflecting the knowledge, imagination and experience of the researcher. The hypothesis can be formulated by understanding the problem, reviewing the literature, visualizing the outcome of the study and considering other factors. Research hypothesis is different from statistical hypothesis, and it is more oriented towards the alternative statistical hypothesis.
- The section may end with the existing gap in knowledge on the topic and justify the need to take up the proposed study.
- *(This section can be written using mixture of past, present and future tenses.)*

19.2.3 Aims and Objectives

The aims and objectives must flow from the justification, background and scope of the study explained previously.

- Aim is a concrete and definite statement highlighting the expected outcome of research on the topic and the objectives are sub-activities having specific relevance and outcome in support of the overall aim.
- While the aim is normally a single statement revealing the ultimate goal of the study, there can be three to four specific objectives each drawn in a sequential manner having a definite outcome in support of aim of the study.
- Objectives are normally statement starting with infinitive form of verbs such as 'to study', 'to assess', 'to examine', 'to explore', 'to relate', 'to correlate', 'to work out', 'to find out', 'to test', 'to search', 'to confirm' and so on.

- The aim is written as a concise statement depicting the ultimate goal/outcome of the study. The objectives are to complement results/sub-outcomes in support of the aim or goal of the study.

19.2.4 Review of Literature

- It is a very important part of a research synopsis which familiarizes the researcher and the reviewers to the problem under study.
- It describes the work done by others at international, national and regional levels on the topic or on similar topics during the recent period.
- It helps the researcher to understand the status of work carried out on related topics and to identify the research gap to set his/her platform for the proposed research work.
- It also helps to know about difficulties faced by other researchers and the corrective steps taken or modifications made by them. The researcher can anticipate similar or additional problems during the study and the review of literature will help him/her in rectifying those problems in advance.
- The review of literature in a synopsis needs to be latest and relevant to the topic. Literature can be reviewed by using various scientific information-gathering methods. These are journals (national or international); publications and websites of research organizations; books; computer-assisted searches; and personal communications with other researchers. The Internet provides a vast opportunity for information gathering.
- Each review begins with the author's name (reference year) and should reflect the aim of the study, location of study, broad research method and design, major outcome, result and conclusion.
- (*This section is normally written in past tense. Normally, five to ten most relevant and recent reviews can serve the purpose in a synopsis.*)

19.2.5 Research Methodology

- In a synopsis, the research methodology forms the core of the research project. The methodology should cover aspects like the study method and design, study settings, target population and sampled population, sampling plans, sample size, inclusion and exclusion criterion, variables under study, data collection, data analysis techniques to be used, ethical clearance, etc. which the researcher intends to apply during the conduct of research.
- It gives a clear picture of the proposed research during the execution part including a tentative data analysis plan, statistical tools to be used, etc.
- (*This Section is generally written in future tense.*)

19.2.6 References

- All references quoted in the introduction, review of literature and anywhere else in the synopsis should be listed here.
- There are various styles for writing references, like Vancouver style, Harvard style, Oxford style and many others. Only one method as prescribed by the degree awarding university or research institute will have to be followed by the researcher consistently.
- *(Note: A synopsis must also contain the information like Name of the researcher and designation, name and designation of the guide, name and designation of head of department/*

institution, name of the institution, signatures of all officials with seal. Generally, every institution has its own format for this purpose.)

19.2.7 Tips for Research Topic Selection and Synopsis Writing

Research: It is a systematic and scientific probe for NEW information, knowledge, concepts, theories, techniques, technologies, inventions, processes, products or solutions to a problem. The academic research has four stages: Planning (including synopsis preparation), execution (data collection), data analysis (subject matter and statistical), report (thesis) writing and publications.

Selection of Research Topic: It is vital for the timely and successful completion of research work. The researcher must select a topic of interest to him/her as well as to the professional society. Emphasis may be given to location-specific and problem-oriented topics. The following steps are suggested:

- Efforts need to be made to select a topic of interest to the researcher which is problem-oriented and location-specific to the extent possible.
- Read and thoroughly understand various concepts and researchable issues of the selected subject matter area from standard textbooks and research journals related to the subject.
- Shortlist a few tentative topics and eliminate those topics recently taken up in the department to avoid duplication of research.
- Based on the FINER approach check the feasibility in terms of availability of cases, time, resources, expertise; interest of the researcher; novelty of the topic; ethical issues and its gravity; and relevance of the research topic in the contemporary context, select a topic for research.
- Discuss the topic with the research supervisor, faculty members and peers in the department to get clearance at the departmental level.
- Once the topic of research is cleared at the departmental level, start an extensive review of research work done on similar topics at national and international levels from the latest to backwards by searching in different journals of repute and by visiting websites of different research institutes.
- Select about seven to eight latest published papers relevant to the identified topics and read it thoroughly to understand the complexities of the topic, methodologies used, type of analysis done, outcome of the research, present stage of research on the topic, research gaps and emerging issues so as to proceed further with the synopsis writing including modifications of the topic, if required.

Synopsis Preparation: The research synopsis is also called research proposal, research protocol, research outline, etc. A well-prepared synopsis is the road map of research and makes the task of the researcher easy at various stages of research. A good synopsis will give maximum information in minimum words. A well-conceived synopsis will go a long way in perceiving the problem by the researcher and convincing the reviewer about the ability of the researcher to conduct the project. In the cases of need for financial assistance, a well-documented synopsis will help to consider the request favourably. Thus, research workers should make all efforts to prepare a well-structured synopsis. A well-thought and well-written synopsis is the crux of a scholarly research project which makes the task easy and systematic for the researcher. Some tips for writing the research synopsis are summarized in Table 19.1

Table 19.1 Tips for Research Topic Selection and Synopsis Writing

S. No.	*Section*	*Tips*
1.	Topic/Title	• Short, precise and self-explanatory, justifying and depicting the specific goal and aim of research topic as finalized based on FINER approach. In any case the title should not be very lengthy.
2.	Introduction	• Definition and background of the problem/origin of the problem, key issues of the selected topic, spatial and temporal gravity/ pattern, current relevance, implications, emerging researchable issues, already researched aspects and outcome by others, research gaps, what is intended to cover by the researcher, research hypothesis and other relevant matters with reference to the topic. • This section should fully justify the topic selected for research. • References, if any, taken from books and other published documents must be cited then and there in brackets and full reference in standard form be given under bibliography/references. • Past, present and future tenses can be used in write-up of this section as per the situation.
3,	Aim and Objectives	• Must directly flow from the problem/justification explained above. • Aim is a single statement highlighting the expected outcome/ research hypothesis (normally in present tense). • Objectives (two to four) which reflect the supporting strategies for achieving expected results and evidence to support the main outcome of proposed research. • Objectives are presented mostly in the infinitive verb form like 'to examine, to explore, to correlate, to study, etc.'.
4.	Methodology	• It includes aspects like study population, sampled population, research design, sample size, inclusion and exclusion criterion, data collection procedure as per subject matter requirement (a flow chart for sequence of action/steps is preferred), broad data analysis plan like estimation of mean values or proportion, standard deviation, correlation coefficients, test of significance (Z-test, *t*-test, ANOVA, chi-square test and other statistical methods to be used for data analysis. • To be written in future tense only.
5.	Review of Literature	• It includes published authored books, published edited books, published papers in journals, websites of research organizations, published seminar/conference proceedings, etc. (online or offline). • Seven to eight latest reviews, both national and international, very much related to the topic of study may be presented in a standard format starting with name(s) of authors, place of study, aim of study, study design, etc. The main findings/conclusion of the study may be summarized in one or two sentences. • The reference should be given in a serial order as per citation sequence under this section in the text of the synopsis. • To be written in past tense only.
6.	References	• The list of references may be given at the end in the required style starting with name of author(s), title of paper, name of journal in standard form, volume with issue number in bracket and page numbers of the journal at the end for each reference. For more clarification, the guide for the style may be downloaded and clearly understood by the researcher for use of punctuation marks and other requirements.

(*Continued*)

Table 19.1 (Continued)

S. No.	*Section*	*Tips*
7.	Data Collection Format/ Performa and Master Sheet	• A data collection format keeping in view the data needed to cover the aim and objectives of the study and to write the results and its discussion with facts and figures in the final thesis. • No unwanted data should be collected, and no data required to cover the objectives should be left out. • A computerized master sheet for data entry may be developed before data collection starts so that the researcher can enter the data on Excel sheet when collected. • A one-to-one correspondence may be ensured between data format and master sheet.

19.3 Writing Research Thesis/Reports

There are several chapters/sections for a research thesis/report which again vary from institution to institution with minor changes in contents and sequences. These are:

- Introduction
- Aim and Objectives
- Review of Literature
- Materials and Methods/Methodology
- Results
- Discussion
- Summary and Conclusion
- Bibliography

The moderate size of a research thesis is a welcome fact. However, there is no strict norm on the size of a thesis as it can vary according to the nature of the problem and also the scope of research including research methods and design. Normally, one-third of the size of a thesis may go for the introduction, review of literature and methodology chapters. The results and discussion may also cover the remaining size almost equally. Wherever the abstract of the thesis is mandatory it may be a one or two-page summary covering importance, objectives, methodology, results and conclusion of the study.

19.3.1 Introduction

It is the enlarged version of the same section given in the synopsis with more facts and figures on the gravity of the selected topic/problem in global, national and regional perspectives. It may include efforts made elsewhere to tackle the problem and the gaps identified from the already existing research. The topic of current research may be thoroughly justified as an attempt to bridge the knowledge/information/process/gap. The aim and objectives should have a perfect match with the justification and importance of the topic selected for the study. The research hypothesis focusing on the major outcome of the research may also form a part of the chapter on introduction. Normally past and present tenses are used in the write-up of this section.

19.3.2 Review of Literature

A classified review of literature based on the major objectives with appropriate sectional headings is preferred. The reviews can be in support of or in contrast to the findings of the study. A synthesized presentation of a review of the literature with a sectional summary at the end of each section will add to the value of this chapter. Normally, a review starts with the author(s) name, year of study brackets, theme of study, location and methodology used in brief, salient findings/outcome of study, etc. Generally, the write-up of this chapter is in past tense.

19.3.3 Research Methodology/Materials and Methods

It should cover the specific research method and research design along with the target population, sampling plan and sample size used for the study. The inclusion and exclusion criteria used to select the ultimate subjects of study must be specified. The procedure applied for information/data gathering from participants including approval of the institutional ethical committee may be indicated. If the study is interventional, the treatment and control may be specified and details as to how the treatment was implemented may be explained. Instruments, if any, used for the research data collection and the details of it including its reliability and validity may be spelt out. Under data analysis, the details of the procedure and techniques used in data analysis including the statistical methods used may be fully explained. The write-up is in past tense.

19.3.4 Results

It is better to present the results of the study in separate sections in accordance with the objectives of the research or research questions. The scientific way of presenting results using well-structured tables, diagrams and graphs along with textual explanations is required. The section-wise highlights of results at the end of each section may be meaningful. This chapter is generally written in past tense.

19.3.5 Discussion

It is better to start with research questions as per the objectives and describe as how the results enable to answer the specific and general questions posed in the study. The inter-objective analysis along with evidence in support or against the outcome of the study may be explained based on studies conducted by other researchers in similar studies by citing proper references. This chapter is also written in past tense.

19.3.6 Summary and Conclusion

Summary is recapitulation of the study results in a concise form. Conclusion is the affirmation of the aim of the study with implication in precise form with recommendation emerging from the study for future purposes. This chapter is generally written in past tense except the recommendations which are generally in future tense.

19.3.7 Limitations of the Study

It is better to spell out the limitations observed/experienced or assumptions made before and during the course of the study as a caution.

19.3.8 Bibliography

A comprehensive list of materials referred to since the formation of the synopsis, discussing the results, to drawing of final conclusion of the thesis may be presented at the end of the thesis in accordance with the selected way of writing the bibliography.

19.3.9 Tips for Writing Thesis

The research thesis is to be prepared based on the plan of work as per approved synopsis only. The writing of thesis gives a practical training to the research scholar to write technical reports and papers which is very important for any professional research degree. Some tips for the preparation of the research thesis are given in Table 19.2.

Table 19.2 Some Tips for Writing the Research Thesis

S. No.	*Sections*	*Tips*
1.	Topic	• Short and self-explanatory, depicting the specific goal and aim of research topic as per approved synopsis.
2.	Introduction	• Definition and background of the problem, origin, spatial and temporal gravity/pattern, current relevance, emerging researchable issues, already researched issues and outcomes by others, research gaps, what is intended to be covered by the researcher, research hypothesis, etc. • This section should justify the topic selected for research. • References, if any, taken from books and other published documents must be cited then and there in brackets and be given under bibliography/references. • Past, present and future tenses can be used in the write-up as the situation warrants. • It may be extended generally to ten to 12 pages.
3,	Aim, Objectives and Research Hypothesis	• Must flow from the problem/justification explained earlier. • Aim is a single statement highlighting the expected outcome/ research hypothesis (normally in present tense). • Objectives reflect strategies for results and evidence to support the main outcome of the research. • Objectives are presented mostly in the infinitive verb form like to examine, to explore, to correlate, to study, etc. • Research hypothesis as a statement depicting the expected conclusion in view of the aim of the study. • It may cover one to two pages.
4.	Methodology	• It includes items like study population, type of research design used, sample size and its calculation formula, inclusion and exclusion criterion, data collection procedure as per subject matter requirement and approved synopsis and a flow chart showing the steps followed in collecting the data/information from respondent is quite welcome. • The scaling and scoring technique, if any used, for quantifying qualitative data may be explained. • Plan of data analysis like estimation of mean values or proportion, standard deviation, confidence interval, coefficient of variation (CV), correlation coefficient, test of significance like Z-test, *t*-test, ANOVA, chi-square test, etc. with formula used may be explained. • To be written in past tense only. It may span six to eight pages.

(*Continued*)

Table 19.2 (Continued)

S. No.	*Sections*	*Tips*
5.	Review of Literature	• Most relevant to the topic and latest publications be included. • It includes published authored books, published edited books, published papers in journals, websites of research organizations, published seminar/conference proceedings, etc. • Thirty to 40 latest reviews closely related to the topic may be presented in a standard format starting with name(s) of authors, place of study, year/period of study, aim of study, study design, etc. The main findings of the study may be summarized in two or three sentences along with conclusion of the study. • The citation reference should be given under bibliography as per required style in serial order as per citation-sequence in the text of the thesis. • To be written in past tense only. • It may cover ten to 15 pages.
6.	Results	• These sections can have sub-sections like profile of selected sample, and all other results as per objectives of the study. • The results are to be presented using tables, diagrams, graphs, pictures, photographs, slides, etc. These are to be serially numbered with proper headings for different categories. The salient findings emerged from each of the tables/diagrams may be explained. • Results are to be explained in text form for each of the table and diagram, etc. • To be generally written in past tense. • It may cover the required number of pages.
7.	Discussion	• The objective wise results are to be discussed with reasons, implications, consequences, etc. • The similarities or divergence observed in results with the studies included in the review of literature section may be highlighted. • To be written in past tense and implication of the study, if any, may be given in present or future tense forms. • It may cover the required number of pages.
8.	Summary and Conclusion	• The whole study is to be summarized section-wise in different paragraphs with the conclusion of the study at the end. • To be written in past tense.
9.	Bibliography	• To be written in the required style. • The list of references may be given serially at the end of the thesis giving name of author(s), title of paper, name of journal in standard form, volume with issue number in brackets and page numbers of the journal at the end for each reference. • Punctuation marks as per required style may be followed.

19.4 Writing Research Papers

Publication of the research paper is a means to share the research outcome, of the study conducted by the researcher, with the scientific community. It is important for the researcher to get recognition for his/her work on one side and for the scientific community to carry forward the research in a continuous manner for the benefit of the subject and society as a whole. Unless the research outcomes are published well in time, the utility of the research will be marginalized, and the researcher's efforts will remain unnoticed. In most of the degree awarding institutions, publication of a specified number of research papers is made mandatory. However, writing a research paper is a cumbersome task.

The text of the observational and experimental articles in most of the disciplines follows the IMRAD structure consisting of an introduction, methods, results and discussion. The title of a scientific paper must reflect the aim and scope of the paper. Additionally, the title must be short but complete. The name(s) of the author(s) and their institutional affiliations also form part of the title page. Most of the journals seek the abstract or summary of the paper giving importance to the study as introduction, objectives, material and methods, location of study, summary of results and conclusions.

19.4.1 Introduction

The section on introduction or importance of the research paper must give a clear definition of the theme of the study and also the issues addressed in the paper. It can include a brief description of global/national/regional significance of the theme of the study, studies conducted elsewhere in a nutshell and the need for the study in its present form. The objectives of the study must emerge from this section and the research hypothesis covered in the paper may also be indicated here.

19.4.2 Materials and Methods

The materials and methods must spell out the research method and design applied for the study and also the way it was performed. The study as quantitative or qualitative, basic or applied, descriptive or analytical, prospective or retrospective may be mentioned. If it is a sample-based study, the actual population sampled, the sample size, method of selecting the sample, the method of data collection used along with the quantitative techniques used for data analysis must be spelt out. If it is an interventional study, the description/profile of the experimental group and control group may be given. The inclusion and exclusion criteria followed for the final selection of the sample must be spelt out.

19.4.3 Results

The results are the outcome of research in quantitative or qualitative form supported with tables, diagrams, graphs, charts, etc.

19.4.4 Discussion

The discussion part of a research paper must be more critical indicating what is special in the outcome of the study as compared to other studies by giving references then and there. The theoretical and practical importance of the findings is to be highlighted.

19.4.5 Conclusion

The conclusion is the most significant outcome of the study which can be used for policy interventions or for other follow-up actions.

19.4.6 References

The references must be written according to the requirements of the journal.

19.4.7 Steps in Writing Research Paper

The task of publication of research begins during or after a research work/project is completed. Generally, a research project is teamwork by a team of researchers from the same discipline or by a multi-disciplinary team. The idea of a research paper can come up from any one member or from a sub-group of the research team. In any case, it is always advantageous to follow a series of steps in writing research papers for early publication.

- **Identify the Author(s):** If the research work is done by any individual, he/she can be the sole author of the research paper. In case any additional work is required to be done with the help of someone else for the proposed paper such persons can also be co-author(s). But, if the research paper is based on a research project carried out by a team of persons, then the author can co-opt other members to form a team of co-authors based on their expected contribution in finalizing the research paper.
- **Planning**: Advance planning for the research paper will enable the authors to include any additional experiment/trial required for the paper to form part of the main project. It will save time and extra work needed for the paper.
- **Title and Abstract**: Fixing a tentative title depicting the proposed content and coverage of the research paper along with a tentative abstract will help to proceed with clarity and time limit.
- **Type of Publication:** Any peer-reviewed journal can have different categories of papers such as: (i) Full-length research articles on IMARD format—introduction, methods, results and discussion, (ii) short communications or (iii) rapid communication. The length/size of each of these categories will be as per norms of the respective journals.
- **Selection of Journal:** The selection of Journal for the publication of a research paper is normally done by adjudging (i) the focus area/subject matter of the journal (basic, theoretical, applied, clinical, etc.), (ii) language (English, local language, Hindi, others), (iii) indexing status of the journal (Scopus, Pub Med, Medline, Google Scholar, Index Copernicus, Open-Gate, etc.), (iv) Availability of the journal (online/hard copy form, etc.), (v) reputation of the journal (based on professional standing of the editorial board, impact factor, acceptance rate, etc.), appearance of a published article (format as font size, font type, etc.), charges for publishing a paper, time taken to publish an accepted paper, quality of figures, etc.
- **Familiarize the Requirements and Norms of Selected Journal:** The authors must carefully read and understand the instructions to the paper writers of the journal related to content, format and other guidelines. It will help to minimize the post submission corrections and changes as well as early publication of the paper.
- **Prepare the Outline of the Paper:** The outline as the roadmap may be prepared before one starts writing the paper and the tasks to complete the paper may be enlisted.
- **Task Distribution Among Authors:** The specific task to be carried out for completion of the paper by the members of the team may be finalized in the beginning.
- **Collection of the Materials:** All related matters for preparation of sections like introduction, methods, results, discussion, references, etc. may be collected and stored in different folders.
- **Prepare the Tables, Figures and Legends:** Prepare all tables, figures, etc. before you start writing the paper.
- **Prepare the First Draft:** As per the task agreed upon by the authors, each author will carry out the assigned task and the first draft of the paper may be prepared. There can be a mixture of styles as well as gaps in the first draft written by different authors which needs unification/correction as per the requirements of the selected journal.

- **Revise the Manuscript:** This can be done by any one author who is well-versed with the content and context of the paper. The gaps will have to be filled and unification as per requirements of the journal may be completed. Wherever restructuring is needed to have a logical flow and order, the same must be done. The required corrections for grammar and spelling will have to be made. Finally, format the document to make the manuscript attractive and easy to read.
- **Check the References:** The citation made in the text will have to be thoroughly checked with references for correctness.
- **Finalize the Paper Title and Abstract:** Based on the final version of the paper the title and abstract can be finalized.

19.4.8 Submission and Revision of Papers

Timely submission of papers in the required format as per guidelines of the journal is very important for early publication. For referred journals, in case the comments of the referee seek revision of the paper, it has to be done within the time limit for resubmission. All comments of the referee will have to be attended meticulously before resubmission within the time limit.

19.4.9 Indexing and Impact Factor of Journals

19.4.9.1 Indexing of the Journal

Indexing of the journal is made as a means for quality assurance of journals. It is the process of listing journals by discipline, type of publication, etc. It is also known as bibliographic indexes or citation indexes. It facilitates easy online searches of articles and acts as a means for information retrieval for libraries. The inclusion of a journal in an indexed group of journals involves rigorous scrutiny and assessment to ensure scholarly publishing standards based on factors like the scope of the journal, its registration under International Standard Serial Number (ISSN), regularity in publishing, competency of the editorial board, peer reviewing of articles before publication, copyright licenses, etc.

19.4.9.2 Impact Factor of a Journal

Impact factor of a journal is an indicator of the relative importance of a journal in a particular field. It reflects the frequency with which an average paper of that journal is cited in a particular time period. Journals listed under the Science Citation Index Expanded (SCIE) or Social Science Citation Index (SSCI) can get an impact factor. The impact factor of a journal can be calculated after the completion of a minimum of three years of publication. The journal with the highest impact factor is the one which published the most cited articles over a period of two years. The impact factor applies to journals and not papers. If a journal has an impact factor of four during 2022 then its papers published in 2020 and 2021 must have received four citations on average in 2022. The impact factor for 2022 can be calculated after the completion of 2022 and will come into force in 2023. The impact factor is calculated as:

Impact factor of a journal during 2022 = A/B where 'A' is the number of times articles published in 2020 and 2021 were cited by indexed journals during 2022 and 'B' is the total number of citable publications like articles and reviews published by the journal during 2020 and 2021.

19.4.9.3 H-Index

H-index is an indicator to represent the publication productivity of a researcher. It is calculated by counting the number of publications for which an author has been cited by other authors at least that same number of times. An author is said to have an h-index 'k' if he has published at least 'k' papers, and each paper has been cited at least 'k' times.

Some Important Points

- Research protocol/synopsis/proposal normally includes sections like title, statement of the problem and hypothesis, aims and objectives, review of literature, research methodology and references.
- There are seven major parts of a research thesis/report: Introduction, review of literature, methodology, results, discussion, summary and conclusion and bibliography.
- The sections in a research paper in most of the disciplines follow the IMRAD structure consisting of introduction, methods, results and discussion.
- The instructions given by the editorial board of the concerned journal must be thoroughly understood and strictly followed for early publication and to avoid repeated revision of the paper.
- Indexing of a journal is made as a means for quality assurance of journals. It is the process of listing journals by discipline, type of publication, etc.
- Impact factor of a journal is an indicator of the relative importance of a journal in a particular field.
- The journal with the highest impact factor is the one that published the most commonly cited articles over a period of two years.

Suggested Readings

Bero, L. A., R. Grilli, J. M. Grimshaw, E. Harvey, A. D. Oxman, and M. A. Thomson, 'Closing the gap between research and practice: an overview of systematic reviews of interventions to promote the implementation of research findings', *British Medical Journal* 317: 465–468, 1998.

Evans, D., and A. Pearson, 'Systematic reviews: gatekeepers of nursing knowledge', *Journal of Clinical Nursing* 10(5): 593–599, 2001.

Mauch, J. E., and J. W. Birch, *Guide to the successful thesis and dissertation: a handbook for students and faculty*, 5th ed., Marcel Dekker, New York, 2003.

20 Work Cited, References and Bibliography

20.1 Introduction

When a thesis is written for the award of a research degree by an institute or a university or one writes a research paper for publication in a standard research journal or a book for publication, the author has to refer to many authentic and published information sources. Usually, such original sources of quotation or paraphrasing are depicted then and there in the text, and it is known as the in-text citation. It is necessary to give a list of such sources in a scientific manner at the end of the thesis or the research paper or a book which enables the readers of the document to go through the referred original sources and also helps the reviewers of the document to verify any points of clarification while assessing the acceptance of the document for its publication or acceptance of thesis for award of degree. The presentation of such sources at the end of the documents is termed as a 'list of works cited' or 'list of references' or bibliography. It is a mandatory requirement to make a document acceptable for academic publication. These terms are sometimes used as synonymous to each other despite some definite differences among them.

In-Text Citation: As far as in-text citation is concerned the major styles include numerical numbers as superscript, numerical numbers in parentheses, author name and year in parentheses, footnote citation at the bottom of the page, author and page number and so on.

List of Work Cited: While writing or composing a research thesis or a research paper the authors have to refer to a number of already published sources by others to substantiate their statements or results either by quoting or by paraphrasing the same. A list of all such sources each starting with the author's name and arranged alphabetically or chronologically is called a list of works cited. Normally a double space is left between each work cited. A hanging indent is used for the first line of each source implying that the first line begins at the left margin of the page and subsequent lines follow uniform spacing for all sources.

List of References: A reference list is similar to a list of works cited meaning that only cited sources in the main document are listed at the end of the document. Standard format for the reference list is followed by different publishing groups. Hence different styles of preparing reference lists are available.

Bibliography: It is a list of all the sources of information or published works consulted by the authors in preparing the document (a thesis, a book or a research paper) irrespective of citation in the text of the document. In general, the elements of a bibliography include the author's name, title of the article, title of the publication, date of publication, the place and name of the publisher, the volume number, issue number and relevant page numbers referred. Different publishers have their own suggested style of writing the bibliography. The comparison of the reference list/list of work cited, and bibliography and features are given in Table 20.1.

DOI: 10.4324/9781003527183-20

Table 20.1 Comparison of Features of Reference List and Bibliography

Features for Comparison	*Reference List/List of Work Cited*	*Bibliography*
Meaning	A list of all sources of publications that have been cited in the text of the document (thesis or research paper).	A list of all sources of publications that have been cited, consulted and gone through for preparation of the document.
Sources	Primary or original.	Both primary and secondary sources.
Context	Used for text citation.	Used for text citation or for conceiving the idea or concept.
Purpose	To support statement/argument/result.	To support statement/argument/result or conceive an idea or concept.
Uses in Types of Documents	Thesis and dissertation.	Journal papers and published research work.

20.2 Purpose of a List of References

A list of references given at the end of the document serves many purposes. Some of these are:

- It substantiates the quantum of the literature review made by the researcher.
- It supports the statements/results of the author.
- It enables the readers to equip themselves with further reading on the topic.
- It is a means to give adequate credit to others who did the actual work and inspired or helped the authors in their research and its documentation.
- It helps to minimize plagiarism.

20.3 Materials for References

The major sources of references (online and offline) in scientific writings include the following:

- Published research journals.
- Published (edited or authored) books.
- Encyclopedias.
- Atlases, dictionaries, directories, handbooks, biographies, yearbooks, etc.
- Websites of institutions.
- Seminar/conference proceedings.
- Magazines and newspaper articles.
- Any other published works.

20.4 Main Parts of a Reference

The main parts of a reference include:

- Authors'/editors' name.
- Title of article/book/publication.
- Name of journal/source in which published/publisher.
- Year of publication.
- Volume and issue number.
- Page numbers covered in the journal.

The sequence, pattern and format are different for various styles of presenting references. Some online journals provide a digital object identifier (DOI) also for each article which is a string of numbers, letters and symbols to uniquely identify an article or document along with a permanent web address (URL). It will help the readers to locate the original document from the reference/citation.

Here it may be important to compare the features in a reference as compared to a bibliography.

- References are the sources that the paper writer has cited in the paper, but the bibliography includes the sources that the researcher has used irrespective of their citation in the document.
- The references and bibliography appear at the end of the documents.
- There are various styles for the text citation of a document.
- A bibliography generally covers information like the author's name, title of the work, name and location of the institutional publisher, date/year of publication and the page numbers of the article in the journal to help readers to track the original material.
- A reference consists of the previous information for the sources cited within the text of the paper.
- The references are arranged chronologically or alphabetically based on the name of the first author.
- The order of giving the information and the punctuation marks followed in different styles are important for both reference and bibliography.
- The citation methods can be different for different journals as per the standard style of citation and writing references and bibliography.

20.5 Different Styles of Writing Citation, References and Bibliography

There are various standard styles to make in-text citations and also to make lists of references or bibliographies for documents like published research papers in journals, theses, books and other such documents. Some of these styles are:

- American Psychological Association (APA style)
- American Medical Association (AMA style)
- Modern Language Association (MLA style)
- Institute for Electrical and Electronics Engineers (IEEE style)
- Harvard Manual of Style
- Vancouver Manual of Style
- Bluebook Manual of Style
- Modern Humanities Research Association (MHRA style)
- Oxford Standard for Citation of Legal Authorities (OSCOLA style)
- Chicago Manual of Style
- American Chemical Society (ACS style), etc.

The information for each reference or bibliography may include author(s), title of article/book, name of Journal/publisher of book, year of publication of book or journal, volume and issue number of the journal, page numbers covered for the referred article, etc. For edited books the chapter title and author(s) of the chapter may also be included. For online journals, the digital object identifier (DOI) for each article which is a string of numbers, letters and symbols to uniquely identify an article or document along with a permanent web address (URL) may also be given. For each style there is a unique way of presenting the citation and also in the sequence of giving required information under references/bibliography, the capitalization of words in the title, name of journal and book font size, font type, line spacing, etc. Some of the main styles and their features for the in-text citation, reference list/bibliography are given next. For each article/document, any one of these styles is to be followed and it is advisable for the author(s) to get familiarized with required style for different sources of citation/reference by going through the online guides available on the internet.

20.5.1 APA Style

It is the style followed by the American Psychological Association. It is used for social sciences and behavioural sciences publications. APA guides are available for the citation method and writing references for the users.

20.5.1.1 Citation Method

APA style follows author/date method of citation where author's last name (generally surname) followed by the year of publication is inserted at the appropriate place in the text of the paper. The references given at the end of the paper must include all the information needed to locate each source.

20.5.1.2 Reference List

The reference list at the end of the paper includes only those references cited in the text of the paper arranged in alphabetical order by the surname of the first author. APA references include information about the author, publication date, title and source. The sequence of information with respect to each reference, the spacing, the punctuation marks, capitalization underlining, etc. is to be followed as per guidelines. The reference materials may include authored and edited books, journal papers and articles, newspaper articles, magazines, microforms, audiovisual media, electronic media, internet sources, etc. The format for writing reference as per APA style:

- Authors' name: Authors' full name beginning with last name followed by first name and middle name with space in between names.
- Publication date: (year, month, date (2023, April 30)).
- Article title: The first word of the title and proper names appearing in titles only are capitalized.
- Periodic title: The title of the periodical is given in italics with all major words in capital letters.
- Volume/issue: When the volume number alone (without issue number) is given, it is in italic form and when the issue number is available, it is given in parentheses after the volume number, both in normal form.
- Page range: The first and last pages of the article in the referred periodical (source) are given as the page range without 'p' or 'pp'.
- DOI: It is the unique digital object identifier for online journal articles, if available.

The APA guidelines available online may be referred for writing citations and making reference lists based on different reference sources like books, periodicals, journal papers and articles, magazine articles, newspaper articles, electronic sources, etc.

20.5.2 AMA Style

The AMA style for citation and referencing was devised by the American Medical Association (AMA). This method is widely used in the field of medical sciences. It follows the reference style starting with the author's last name and initials, source title, information about the publisher followed by publication date. No indent is required if any list of references extends to the second and subsequent lines. There are specific guidelines for different types of sources.

20.5.2.1 Citation Method

The in-text citation under AMA style is made by using superscript numerals where the number is the serial number in the list of references given at the end of the document. The reference number should be given after the fact or quotation, or idea cited in the text.

Example: Significant differences in the level of immunization of infants according to their gender were reported by Varghese et al.[5]—when the article on gender effect on immunization by Varghese and others stands cited as the 5th one in the order of sequence of citation in the text made by someone on immunization of children. When there are multiple consecutive sources (say references two to four) to be cited at one place it may be cited as[2–4] at the appropriate place in the document. When the references of a cited statement in the document are not consecutive it may be mentioned with serial numbers of reference sources separated with commas like[4,6,8].

20.5.2.2 Reference List

A numbered reference list is given at the end of the document. The reference list must include all the cited sources in the sequential order of citation in the text meaning the list starts with the first citation in the text. The purpose of references here is to acknowledge the author who has done the original work and to facilitate the readers with more information on related work. In papers having multiple authors, all the authors are responsible for having the combined reference list as per stipulated norms. Only primary sources and references fully read by the authors are to be included. Each reference starts with the author's surname followed by initials without periods. When there are six or less number of authors, all names are to be mentioned. If there are more than six authors, the first three are individually mentioned followed by 'et al.' The title of the article is to be retained as appears in the original title (spelling, abbreviation, capitalization of words, numbers, etc.). The name of the journal as per the National Library of Medicine (NLM) abbreviation may be used for different journals. The sequence for each reference is author(s), title of paper, journal name, year, volume (issue) and page numbers. The reference list may include journal articles including online (with URL, with DOI and published ahead of print), book chapters (print and online), books (online and print), websites, monographs, government/organization reports (print and online), patents, materials accepted for publication and conference presentations (print and online). All the references are numbered numerically in the order in which they appear in the citation of the document.

For more details about the use of capitalization, punctuation marks, font type, etc. the AMA Manual may be referred.

20.5.3 MLA Style

This method of citation and referencing was developed by the Modern Language Association. It gives details for text citation in the body of the paper and preparation of a list of 'Works Cited' at the end of the document. This method of citation and referencing is widely used in the arts and humanities.

20.5.3.1 Citation Method

The citation in the text is made by giving the surname of the first author followed by the page number in parenthesis without any punctuation between the two.

20.5.3.2 Works Cited List

The reference list is given under the heading 'Works Cited'. It is a list of all sources from where the ideas and information have been taken to prepare the manuscript. The list is prepared in alphabetical order by the surname of authors in sequence. The material in each reference will include the author (surname), title of source (paper/article), title of container (name of journal/newspaper, other contributors (editor/director), version (edition), number (volume/issue), publisher (agency responsible for publication), publication date, location (location or page number). The work cited may include books, chapters in book, journal articles, films, DVDs, video recordings, websites, artworks, illustrations, etc. For more details, the MLA Quick Guide may be referred.

20.5.4 IEEE Style

IEEE style is followed mainly in Electronics, Electrical Engineering and Computer Sciences disciplines. There are modified versions of IEEE such as IEEE-Pervasive Comp, IEEE Micro and IEEE ACM Trans Network.

20.5.4.1 Citation Method

Citation is made when writing academic documents to acknowledge any sources used by the writer. IEEE citation style is numeric with numbers in parenthesis as per references numbered at the end of the document. The citation number is given in square brackets at the appropriate place in the text immediately after the cited matter with a space but without any punctuation. Once a source is cited with a number the same number is to be used at subsequent places where the same source is used.

20.5.4.2 Reference List

The references are given in numeric order at the end of the text in the same order those are cited in the text as per IEEE formatting guidelines. As per these guidelines, the list should be formatted by aligning references left, using single space for each entry and double space between entries. The reference number is given in square brackets at the left margin. The reference materials may include books, book chapters, electronic books, journal articles, E-journal articles, conference papers, reports, patents, standards, thesis/dissertations, datasheets, online documents, websites, etc. The IEEE guidelines for each of these may be followed while composing the references. The sequence of the IEEE reference list is: Author initials, last name, the title of the source, place/city, publisher, publication date and DOI.

20.5.5 Harvard Style

It is one of the oldest referencing systems developed by Edward Laurens Mark in 1881 who was a zoologist at Harvard University. The guidelines for the Harvard style of referencing are available for both in-text citations and full references at the end of the document. It is a widely accepted style of referencing used across many subjects.

20.5.5.1 Citation Method

The in-text citation in this method follows the author-date system which includes the author's name and publication date of sources referred by the researcher. The in-text citations with only one author include author(s)/editor(s) name, year of publication and page number(s). When there are two or three authors the surnames of all authors with year of publication and page number(s) are to be given under the in-text citation. However, when there are four or more authors, the first author's surname followed by 'et al.' may be used.

20.5.5.2 Reference List

The reference list covers all sources referred to while preparing the document showing information like author, date of publication, title, etc. on separate page(s) at the end. It must be organized alphabetically by author or source title when there is no specific author. When multiple works of the same author are cited, then the list is ordered by dates. Each reference must be double-spaced with a blank space between the lines. All in-text citations must be given under reference which may include books, edited books, chapters in the edited book, CD-ROMS, e-mails, interviews, journal articles, newspaper articles, videos, DVDs, Worldwide webs, etc. The online guide for the reference list and in-text citing as per Harvard style may be used for each of these sources.

20.5.6 Vancouver Style

It was developed in Vancouver (Canada) in 1978 by the editors of medical journals. The International Committee of Medical Journal Editors (ICMJE) recommendations include guidelines for the preparation and formatting of the manuscript for submission to biomedical journals for publication. These guidelines are followed by thousands of biomedical journals across the world.

20.5.6.1 Citations in the Text

In-text citation is made using numerical numbers in the parenthesis which are placed after the relevant part of the sentence. The same reference number is used for repeated citation of the reference. Based on the sequences of citation, the references are numbered serially. Place same reference number in parentheses throughout the text, tables and legends of a document.

20.5.6.2 References List

The list of reference is prepared and given at the end of the document based on the sequence of citation numbers in the text matter. Some of the important points related to the reference list as per Vancouver style are as follows:

- Each reference comes only once even when the same is cited a number of times in the text.
- Spacing within reference is single and between references is double.
- For research paper in journals, the sequence of writing the reference is name of author(s), title of paper, name of journal (the name of journal may be given in standard abbreviated form of the name of journal) followed by a period and space, year of publication with a semicolon, volume and issue number (in parentheses) followed by colon and page range of paper with period (like if page number is from 241 to 248 it is written as 241–8).

- For books, the sequence of writing the reference is name of author(s), title of book, edition, place of publication, name of publisher and year of publication.
- In the case of edited books, editor's name may be given in place of author's name followed by word editor(s).
- For edited book with chapters written by different authors, the sequence of writing the reference is name of author(s) of referred chapter, chapter title followed by 'In:', editor(s) name, comma, space, word editor(s), period, book title followed by period, edition number followed by period, place of publication with colon, publisher, if copyright, put 'c' and year of publication without space else year of publication, period and page range of the referred chapter.
- **Pattern of writing name of author(s):** Last name of each author appears first, followed by space then initials without space and period, put comma and space between the names up to six authors and a period at the end of last author. If number of authors exceeds six, 'et al.' is used after the sixth author. In case of edited books, name of editors followed comma, space and word editor(s) in the sequence.
- **Patter of writing title:** Capitalize only the first letter of the first word of the title and remaining part of the title is written in lower case except proper name. If the book is a revised edition, it is shown after the title as 'k^{th} ed'.
- **Patter of writing publication information about book:** After the title of book (edition if applicable), place a period and space and city name of publication with colon. Give the name of publisher followed by semicolon. Give year of publication followed by a period. If publication has date of copyright, give the year of copyright preceded by letter 'c'.

Following are the examples for writing references for authored and edited books and journals which are commonly used by the researchers:

Book:

Daniel WM. Biostatistics: basic concepts and methodology for the health sciences 9th ed. New Delhi: John Wiley & Sons, Inc; c2010.

Edited book:

Upadhyay R, Solanki D. Women empowerment in Thar desert region. In: Varghese Nisha, Burark SS, Varghese KA, editors. Natural resource management in Thar desert region of Rajasthan. Switzerland: Springer Nature; c2023. p 235–47.

Journal:

Srivastava AK, Sud UC, Chandra H. Small area estimation-an application to national sample survey data. J. Ind. Soc. Agril. Statist. 2007; 61(2): 249–54.

The reference material may also include electronic journal article, electronic books, dictionary, online references, Wiki entireties, newspaper articles (paper and electronics), website pages, streaming videos, electronic images, poster presentation in conferences, etc. For writing references related to any of the aforementioned sources, it is suggested to refer the online guides available for writing references according to Vancouver style.

20.5.7 Bluebook Style

The Bluebook style is used in the American legal profession. This style is followed mostly in legal writings in the United States by legal professionals. As far back as 1925 it was only an eight-page booklet for Harvard Law and now it is a three-volume manual. It is designed in such a way that the reader can easily find the cited sources.

20.5.7.1 Citation Method

The footnote citation of sources at the bottom of pages may include author(s), title, publisher, year, page/section number.

20.5.7.2 Reference List

The Bluebook reference list is known as the Bluebook bibliography which is an alphabetically prepared list of all cited sources with the author's name or in its absence by the initial word of the title in a legal document. Every entry in the list may cover the author(s) as appeared in original source, title in sentence case formatting with initial word and proper nouns capitalized, publication information like publisher, place of publication, date of publishing, volume, issue number, page number, etc.

20.5.8 MHRA Style

It was developed by the Modern Humanities Research Association for use in the arts and humanities.

20.5.8.1 Citation Method

According to this style, the sources are cited in foot notes of corresponding pages of the text and marked by superscript numbers in the text.

20.5.8.2 Reference List

The reference list at the end of the text includes all sources alphabetically in order by authors' last name.

20.5.9 OSCOLA Style

The Oxford Standard Citation of Legal Authorities (OSCOLA) is a reference and citation style developed at Oxford University.

20.5.9.1 Citation Method

OSCOLA follows a footnote citation system. At the end of a sentence seeking a citation a numeric number is given as superscript after punctuation.

20.5.9.2 Reference List

The complete references corresponding to all superscript numeric citations appearing on a page are given at the bottom of those pages as a footnote.

20.5.9.3 Bibliography

The bibliography is given at the end of the document for all sources cited and referred to on different pages. This list is prepared by type of sources and alphabetically based on the first name of the author under each type.

20.5.10 Chicago Style

Chicago style is mostly used in the humanities and sciences.

20.5.10.1 Citation Method

Chicago style has an author-date way of citation in the text. Author(s) surname and year of publication are given in brackets at appropriate places in the text irrespective of the source of referencing. Another way of citation is 'notes and bibliography' and in this style, a superscript number is given at the appropriate place in the text corresponding to the serial number in the bibliography. This style is practiced in literature, history, arts, etc. whereas author-date style is followed in sciences and social sciences. All the in-text citations must be included in the reference list. For in-text citations, a uniform pattern is followed irrespective of sources of citation as books, journal articles, internet documents, etc.

20.5.10.2 Reference List

The full details of the sources of in-text citations required to search those are given at the end of the document as a reference list. The reference list is prepared alphabetically by surname of the first author.

20.5.11 American Chemical Society (ACS) Style

20.5.11.1 Citation Method

The ACS style of in-text citation follows three methods—by superscript numbers, by numbers in parentheses and by author name and year of publication.

20.5.11.2 Reference List

The reference list under ACS style is given at the end of the document according to the order those appear in the in-text citation. If a reference is cited more than once in the text, the same number is used throughout the text.

Some Important Points

- The way of showing borrowed or perceived facts and ideas from other publications while preparing the text of a scientific document then and there by author(s) is called in-text citation.
- List of works cited or a reference list is the list of all publications and sources that the author(s) used to prepare the document and cited somewhere in the text of the document.
- Bibliography: It is the list of all the sources of information or published works consulted by the authors for preparing the document (a thesis or a research paper) irrespective of citation in the text of the document.
- There are various standard styles for different disciplines/subjects to make in-text citations and to make lists of references or bibliographies to documents like published research papers in journals, thesis and other such documents. These include styles like American Psychological Association (APA), American Medical Association (AMA), Modern Language Association (MLA), Institute for Electrical and Electronics Engineers (IEEE), Harvard, Vancouver,

Bluebook, Modern Humanities Research Association (MHRA), Oxford Standard for Citation of Legal Authorities (OSCOLA), Chicago style, American Chemical Society (ACS) style, etc.

Suggested Readings

Alves dos Santos, E., S. Peroni, and M. L. Mucheroni, 'An analysis of citing and referencing habits across all scholarly disciplines: approaches and trends in bibliographic references and citing practices', *Journal of Documentation* 67(6), https://libguides.reading.ac.uk

Pears, R., and G. Shields, *Cite them right: the essential referencing guide*, 12th ed., Palgrave Macmillan, 2022.

University of Reading Lib Guides, 'Different styles and systems of referencing—citing references', https://libguides.reading.ac.uk

Williams, R. B., 'Citation systems in the biosciences: a history, classification and descriptive terminology', *Journal of Documentation* 67(6): 995–1014, 2011.

21 Ethics in Research and Publications

21.1 Introduction

Ethics implies a set of rules applicable to self and others for formation of a better society. Ethics relates to the value system of our society that covers the consensual agreement on what is right and what is wrong. It is much more than what is legislatively defined as legal and illegal. The ethical principles are important to maintain harmony within and between all sections of society. The scientific community will also have to address and resolve ethical problems in their efforts to make a better society. They, as a group, use members of other social groups to arrive at new knowledge and information, especially in activities like research in different areas. In order to avoid any confrontation between these groups, adherence to ethical norms is vital. An indifferent attitude to these problems may result in crossing ethical barriers, during the phases of implementation of such activities.

Ethical codes are pre-conceived ideas for the behaviour of people as individuals and as a society while doing activities for the benefit of the society. They are moral statements that can be applied to particular situations to help us make decisions and guide our behaviors. Those are linked to cultural values at a defined time in our history and are subject to change as attitudes and values change. What is well thought out to be insensitive today can be normative, just a half-century ago or after. In doing research there may be a conflict between the speedy conduct of a study by the researcher and the trouble of doing what is deferential to humans or even animals. Research ethics include guidelines for the conduct and dissemination of research results. When research is based on human subjects, relevance of ethics is more vital. The ethical issues in research could be perceived as:

- Ethics in conduct of research.
- Ethics in writing research papers.
- Ethics in publication.

21.2 Ethics and Research

The objectives of research ethics are:

- To protect the dignity, rights and welfare of human participants in research.
- To confirm that research is for human welfare.
- To ensure ethical reliability in terms of risk minimization, protection of personal privacy and adherence to informed consent from participants.

Ethical behaviour in a scientific activity is defined as a set of moral regulations, principles, rules or standards governing a person or a professional group conducting that activity. Ethical

DOI: 10.4324/9781003527183-21

principles of conduct in research may include moral regulations related to privacy of data/information, safety of the participants, anonymity of participants, confidentiality of data, avoidance of bias at all stages, respect for respondents, maximization of benefits to participants etc. in addition to informed pre-consent of participants for voluntary participation and freedom to withdraw from the proposed research at any stage without any ill will from the researchers. Besides, the data must be interpreted honestly without distortion by the researcher. The participants must have a due share of ownership and any benefits accruing from the research.

Researchers are focused on knowledge expansion and on the methodology of their projects including personnel and equipment, statistical analysis, selection of subjects, research protocols, sample size and many other technical aspects. At the same time, they must try as much as possible to respect the research environment, which requires attention not only to physical resources including funds, but also to animals or dignity of human subjects used as subjects of study.

Ethical considerations in research may help to decide whether a particular research is to be done, and if it is so, how it should be pursued. Thus, it is vital to be reflective, transparent, sincere and adhere to ethical guidelines in regard to research subjects.

Research involves the systematic process of collecting and analyzing data to increase our understanding of the phenomenon under study. It is the duty of the researcher to contribute to the understanding of the phenomenon and to communicate that understanding to the benefit of the society.

Ethics refers to both morals and beliefs, 'beliefs about what is right and what is wrong'. Ethical issues in research can be raised at all phases of research including the problem definition, stating research objectives/hypotheses, literature review, choice of research design, questionnaire design, data collection procedures, data editing and cleaning, choice of statistical methods, data analysis, conclusions and recommendations, and even referencing. Some cases of unethical research are often associated with particular research methods, such as disguised observation and deception in experiments. Hence ethics is relevant at every stage of the research.

Ethics is now an essential part of any research project and is just another stage of research seeking adherence to ethical norms. Nowadays, doing ethical research is essential to producing relevant and successful findings. As such, researchers' ethical conduct is currently being scrutinized like never before. In today's society, any concerns regarding ethical practices will negatively influence attitudes about science, and the abuses committed by a few are often the ones that receive widespread publicity. Clearly, researchers have liabilities to their line of work, patrons and respondents and are obliged to have high ethical standards to make certain that both the purpose and the information are not brought into ill repute.

As a branch of philosophy, ethics deals with the dynamic of decision-making concerning what is right and what is wrong. Research ethics includes requirements on daily work, the protection of the dignity of subjects and information in the research that is being made known.

A researcher can be a research scholar, psychologist, educationist, economist, sociologist, medical doctor or anthropologist. His primary responsibility is to help protect participants and the aim should be clear. Wherever the informed consent ought to be obtained, protection of participants from any sort of harm must be ensured and privacy of information should be maintained. Some of these concepts are discussed.

Informed Consent: It is a procedure for selected participants to decide in getting involved or not in a study being taken up by the researcher. In order to take a free decision by the participant, the researcher will have to brief him or her in local language all the details regarding aim and objectives, the research method to be used and all possible risk to the participants by joining in the study. The freedom of the participants to leave from the study at any stage also may be made clear to them. The written consent is taken after the participants are well informed all the aforementioned details and hence it is called informed consent. If the participant is a major and mentally alert, the consent taken from him or her is direct consent. If the participant is a dependent child, a minor or an adult not capable of understanding the details of the study, the consent is taken from the guardian or a responsible attendant of the dependent person selected for the study. The consent is called substitute consent.

Harm Minimization: Researchers are required to take reasonable precautions to reduce injury when it is predictable and inevitable and to avoid injuring their patients, students, supervisees, study participants, organizational clients and anyone with whom they work. Upon learning that a participant has been hurt by study practices, the researchers must take appropriate action to reduce the harm. The researcher will have to ensure that the participants of the study or anyone associated with the study are safe and do not receive any harm which includes physical pain, psychological stress, mental strain or any such adversities. While dealing with children, elderly people and disabled persons, the researcher will have to be extra cautious.

Privacy: The privacy of private information and data collected from the study participants which can cause family or social problems to them at any stage will have to be ensured by the researcher.

Deception: Deception of any type by the researcher to the participants is an unethical act. It can be by way of giving wrong information, by hiding actual information from participants and it can happen through omission or commission by the researchers. Informed consent is remedy to overcome any such issues in research.

21.3 Ethics in Quantitative Research

Quantitative research mostly includes experimental or observational research in which data is collected in numerical form either by measuring or by counting. The protection of participants in experimental group who receive interventions is of prime importance to the researcher as it can sometimes cause problems to the participants in the experimental group. On the contrary, if the interventions make positive impacts on those who receive, it is argued that the participants in control or placebo group are deprived of the benefits.

Even though non-experimental designs, such as surveyor interview techniques, may involve less complexity or risk than experimental investigations, it is nonetheless crucial for investigators to understand fundamental ethical guidelines for participant protection, such as 'full disclosure and consent'. For instance, in survey research, all participants should be fully informed about the study's objectives, participant demographics, response confidentiality, intended use of the findings and data access.

21.4 Ethics in Qualitative Research

Qualitative research uses narrative descriptions. Rather than controlling situations, a researcher uses qualitative methods to observe and describe them. The following are fundamental ethical guidelines for qualitative researchers:

- Don't tamper with the study's natural environment. More importantly, participant and non-participant observations are essential to qualitative research and are widely employed in the domains of sociology, anthropology and education. But each raises different moral questions about deceit, consent and privacy.

- Privacy of information supplied by the study participants in group discussion and other qualitative methods is a major problem. The participants in research based on qualitative methods may face problems like criticism, ill-will, counter arguments, etc. from family members and society as well.

The quantitative and qualitative approaches differ from one another; neither is better than the other; rather, both have acknowledged advantages and disadvantages and are best applied in tandem. Identifying and trying to make sense of the conflict that exists between researchers about quantitative and qualitative research might help to generate new and distinct lines of inquiry. The ethical issues in research consist of two major aspects. Firstly, ethical issues related to the conduct of research and secondly, ethical issues pertaining to publications.

21.5 Ethical Issues in Conduct of Medical Research

In the process of research, the researcher may use interventions. All the interventions made must be evaluated through research for its effectiveness, efficiency, accessibility, safety and quality considerations. Therefore, ethical standards that ensure respect for human beings and animals and protection of their health and rights assume great significance. All research is meant for the generation of new processes or products that lead to enhancement in quality of life of the people. As most of the human health and veterinary research is based on interventions made on human beings or animals and hence such researchers must follow these principles:

- Protection of dignity, integrity, life, health and right to self-determination of each person/animal.
- Privacy and confidentiality of all personal information.
- Biomedical researchers must consider the regulatory, legal and ethical norms and standards for research based on human beings and animals.
- Biomedical researchers must ensure minimum harm to the environment.
- Research on human beings and animals must be carried out with full adherence to ethical and scientific norms.
- A physician combining medical research with medical care of any person must ensure that the participation of the patient would in no way adversely affect his/her health.
- Research at the cost of risks and burdens of human subjects must be avoided.
- Risks and burdens of human subjects as a result of his/her participation in research must be continuously monitored by the researcher.

21.5.1 World Medical Association Declaration

The Declaration of Helsinki, which includes ethical principles for research involving human beings and on identifiable human material and data, was developed by the World Medical Association (WMA) as a statement of ethical standards for medical research involving human beings. The declaration is primarily targeted to physicians, in line with the WMA's mandate. The WMA calls on other parties engaged in medical research involving humans to follow these guidelines. The International Code of Medical Ethics states that 'A physician shall act in the patient's best interest when providing medical care', and the WMA's Declaration of Geneva obligates doctors by stating that 'The health of my patient will be my first consideration'. The responsibility of the doctor is to advance and protect the rights, wellbeing and health of patients, including those involved in medical research.

Medical researchers have an obligation to safeguard patients' lives, health, integrity, dignity and right to self-determination. Even when they have granted agreement, the onus of protecting study participants must always be on the doctor or other healthcare providers, not the subjects themselves. When doing research with human subjects, doctors must consider both applicable national and international standards and guidelines. No ethical, legal, or regulatory obligation, whether national or international, shall lessen or eliminate the ethical considerations stated in the WMA Declaration. Medical research should be carried out in a way that limits potential environmental damage. Only those with the necessary ethical standards, scientific knowledge, training and credentials should undertake medical research on human subjects. Research involving both healthy and sick participants must be carried out under the guidance of licensed medical experts. When a study has the potential to be preventive, diagnostic or therapeutic, doctors who integrate medical research with patient care should only consent to have their patients participate in it if they have good reason to believe that their participation won't have a detrimental effect on their health. It is essential to provide fair compensation and treatment to participants who get harmed as a result of their participation in the study.

21.5.2 Risks and Benefits

In medical practice and research, risks are associated with most procedures. In medical research, human subjects are only employed when the advantages of the study justify the risks and difficulties they face. A thorough assessment of the known hazards to the persons and groups participating in the study, as well as the known benefits to them and other individuals or groups who may be impacted by the condition under examination, must precede any medical research involving human subjects. It is important to reduce the risks involved. The risks need to be routinely observed, assessed and documented by the researcher. Doctors are not allowed to take part in a study that uses humans as subjects unless they are sure of handling and managing the associated risks. Physicians must decide whether to continue, amend, or abruptly halt the study when it is determined that the hazards outweigh the possible benefits or when there is undisputed evidence of definitive outcomes.

21.5.3 Vulnerable Groups

Some people and groups are more susceptible than others in getting harmed or injured during an intervention. Such vulnerable groups and people should be given special consideration for protection. Research on vulnerable groups can be justified only when it is not relevant for non-vulnerable groups and is in response to the health needs or goals of the vulnerable group is it justified. This group should also stand to gain from any knowledge, procedures, or treatments that come out of the research.

21.5.4 Scientific Requirements and Protocols

Medical research involving human beings as subjects must adhere to commonly recognized scientific standards and should be supported by competent laboratory and, wherever necessary, animal testing, as well as a complete understanding of the scientific literature and other pertinent sources of information. Respect must be shown for the wellbeing of animals used in the research. A research protocol outlining and justifying the investigation's design and conduct is required for every study involving human beings.

- An explanation of the ethical issues raised and an indication of how the declaration's tenets have been taken into account should be included in the protocol.

- The protocol should also include information on funding, sponsors, institutional ties, any conflicts of interest, subject incentives and procedures for treating and/or compensating study participants who sustain injuries as a result of taking part in the research. The protocol for clinical trials must also provide suitable preparations for post-trial provisions.

21.5.5 Institutional Ethics Committees (IEC)

Prior to beginning of any study in medical sciences, the research protocol needs to be submitted to the appropriate research ethics committee for assessment, approval, comments and guidance. This committee needs to function transparently and be independent of the sponsor, the researcher and any other unethical pressures. Along with any relevant international norms and standards, the rules and regulations of the country/countries in which the study will be done must also be considered.

The committee must have the authority to keep an eye on ongoing research projects. The committee must receive monitoring data from the researcher, including details regarding any significant adverse events. The committee must review and approve any changes made to the protocol before they may be implemented. Following the completion of the study, the investigator is required to present a final report to the committee that includes an overview of the results and recommendations.

21.5.6 Maintaining Privacy of Data

In medical research, prime importance should be given to maintaining the confidentiality of personal details of the participants in the research and to protect their identity.

21.5.7 Informed Consent

In medical research, it is crucial that participants can give informed consent to participate voluntarily. No one who can give informed consent may be involved in a study unless they willingly agree, yet it is also appropriate to consult with relatives or local authorities.

All participants who are able to give informed consent to participate in medical research involving human subjects must be fully informed about the goals of the study, the methods, the sources of funding, any potential conflicts of interest, the researcher's institutional affiliations, the expected benefits and risks, any potential discomfort, the post-study provisions and any other relevant study details. The prospective participant must be informed of their freedom to refuse to take part in the study or to withdraw consent at any time without having to face any consequences. Consideration should be given to the ways in which the information is delivered, as well as the particular information requirements of each possible subject.

Once it is ascertained that the subject has understood the information conveyed, the physician or any other trained person must then obtain the informed consent of the subject, preferably in writing. The non-written consent needs to be officially recorded and witnessed if it cannot be communicated in writing. Every participant in medical research should have the choice to learn about the overall goal and findings of the investigation. A physician must use extra caution when obtaining informed permission from a potential subject for research study participation if the individual is in a dependent relationship with the physician or may consent under coercion. In these circumstances, a suitably qualified person who is totally unrelated to this relationship must request informed consent.

Research involving participants incapable of providing informed permission due to medical or mental conditions (e.g., unconscious patients) may only be conducted if the research group includes members who are required to have the relevant physical or mental condition. In these situations, the doctor has to ask the legally appointed person for informed consent.

The doctor must thoroughly explain to the patient any aspects of their care that are related to the research. A patient's refusal to participate in research or to opt out should never have a detrimental effect on the patient-physician relationship.

Before gathering, keeping, or exploiting identifiable human material or data for medical research, doctors must have informed consent. Research using information or materials from biobanks or other similar repositories is included in this. Obtaining consent for this sort of study may not be possible or viable in some extremely rare cases. Under these conditions, research can only be carried out with the approval and assessment of a research ethics committee.

21.6 Principles in Writing Scientific Papers

The text of research articles based on observational and experimental research follows the IMRAD structure—Introduction, methods, results and discussion. Every research paper must have an informative title, a section on introduction, objectives, methods used in the study, results and discussion, followed by conclusion and a list of references at the end.

21.6.1 Publishing the Results

Regarding the publication and distribution of research findings, there are ethical responsibilities on publishers, editors, sponsors, writers and researchers alike. Researchers are responsible for the accuracy and completeness of their findings and have an obligation to make the results of their work with human subjects available to the public. Everyone involved should follow recognized standards for moral reporting. Both good and inconclusive results need to be disclosed or otherwise made available to the public. Publication requirements include disclosing funding sources, institutional affiliations and conflicts of interest.

21.7 Using Unproven Interventions in Practice

In a few rare situations, a physician may use unproven interventions with the informed consent of the patient or legally authorized representative when he/she is confident that it can save a patient's life, restore his/her health or reduce their suffering. The effectiveness and safety of this treatment should then be investigated through research. Every new piece of information must be recorded and, when appropriate, made public.

21.8 Publication Ethics

The success of academic publishing largely hinges on trust. Readers trust the peer-review process; authors trust editors to choose qualified peer reviewers; and editors trust peer reviewers to offer unbiased assessments. Strong intellectual, commercial and occasionally political interests are present in the academic publishing environment, and these interests may conflict or compete with one another. A sustainable and effective publishing system will be fostered by wise decisions and robust editorial processes intended to manage conflicting interests. Academic societies, journal editors, authors, research funders, readers and publishers will all gain from this system. Ethical publication practices do not emerge by accident; they must be actively pushed in order to gain traction. The following are the fundamentals of publication ethics:

21.8.1 Transparency

Funding sources for studies and publications must always be made available. This should be made clear in editors' editing policies. All publications that authors prepare for publication should routinely include information regarding research funding. When one is able, a clinical trial registration number ought to be mentioned.

21.8.2 Authorship Acknowledgement

According to The International Committee of Medical Journal Editors (ICMJE) authorship criteria, authorship credit should be based on:

- Making significant contributions to the article's conceptualization, design, or data collection, analysis and interpretation;
- Drafting the work or critically editing it to remove any significant intellectual material; and
- Final approval of the version that will be published.

When writing a research paper, authors should indicate if they had full access to the study data used to support the publication. Not all contributors can be considered writers, but they should still be listed in the acknowledgement and their specific roles explained. When gathering authorship data for research papers, the choice of authorship ought to be made at the outset of the investigation. Editors are not responsible for deciding on authorship. Editors ought to require clear and comprehensive credits for all authors who have contributed to a work.

Editors ought to implement suitable mechanisms for notifying contributors about authorship standards (if applicable) and/or acquiring precise data regarding individual contributions. Editors should request that authors include with their initial submission package a statement attesting to the following: that all persons listed as authors have fulfilled the necessary requirements for authorship, that no one who is eligible for authorship has been left off the list, that contributors and funding sources have been duly acknowledged, and that contributors and authors have approved the acknowledgement of their contribution.

21.8.3 Attributing Authorship to a Group

In cases where many authors report on behalf of a larger group of investigators, the International Committee of Medical Journal Editors suggest that 'the group should identify the individuals who accept direct responsibility for the manuscript when the work has been conducted by a large, multi-center group'. These people ought to fulfil all of the requirements for authorship listed above. When submitting a work with many authors, the lead author must specify the preferred citation style and explicitly include each individual author along with the group name. The members of the wider authorship group should be listed as an appendix to the acknowledgement of the individual writers who take direct responsibility for the manuscript.

21.8.4 Protecting Research Subjects

Journals ought to request that authors certify that the institutional review board or research Ethics committee in question has authorized the work they are submitting. Manuscripts involving human subjects must state that the experiments were conducted with the participant's knowledge and proper informed permission, if any.

In cases where there is uncertainty about whether the proper procedures have been followed, editors should retain the authority to reject publications. When a study is submitted from a nation without an institutional review board, ethics committee, or other comparable review and approval process, editors should rely on their own judgement when determining whether or not to publish the work. If it is decided to publish a manuscript in these circumstances, a brief explanation should be provided.

21.9 Misconduct in Scientific Publishing

Research misconduct includes fraudulent and deceitful practices used by researchers in the conduct of research or publication. The most severe forms of scientific fraud include:

- Data fabrication (inventing data/results).
- Data falsification (changing the results).
- Plagiarism.

21.9.1 Fabrication and Falsification

The accuracy of the data is a prerequisite for the validity of the research. Falsification and fabrication are significant problems in scientific ethics because they cast doubt on the accuracy of data. The act of removing or changing research tools, information, or procedures so that the findings are no longer accurately represented in the research record is known as falsification. The act of fabricating data or findings and then documenting, reporting, or recording them in the scientific record is known as fabrication. These two acts are considered as the most serious transgressions as they cast doubt on the integrity of every single component of scientific research.

21.9.2 Plagiarism

The word plagiarism originates from the Latin word plagium which means kidnapping a man. Literally speaking, it is stealing or presenting something that was written by someone else as your own.

Direct plagiarism (plagiarism of the text); mosaic plagiarism (taking concepts and viewpoints from the original source and using a few exact words or phrases without giving credit to the author); and self-plagiarism (reusing one's own work without citations) are three main categories of plagiarism.

Researchers need to use their knowledge and discernment in order to interpret the published data judiciously, integrate prior knowledge into new papers and distinguish between novel concepts and results from previously published papers. Authors must adhere to the moral, legal and ethical standards that the scientific community accepts. They must quote taken phrases and ideas from published or unpublished works, correctly citing pertinent publications. To put it simply, a passage taken verbatim from another author's work must be put in quotation marks or inverted commas.

21.9.3 Multiple Publication

Republication of prior published work is generally not allowed however, reprinting of translated content from an original publication in a foreign language may be accepted by journals, meaning that they will not be considered 'redundant'. Such journals need to ensure that prior permissions are taken and should clearly mention the source of the original publication and it should be made clear to the readers that the material is a translated reprint. It is more difficult to justify republishing in the original language with the intention of reaching a different audience when the original publication is electronic and thus easily accessible; however, editors should take the same actions as for translation if they believe this is appropriate.

21.9.4 Organizations for Prevention of Misconduct

The Committee on Publication Ethics (COPE) is a body. Since its founding in 1997 by a group of UK medical journal editors, COPE has grown to include nearly 7000 members from all academic disciplines worldwide. Academic journal editors and anybody else with an interest in publication ethics can join. Elsevier, Wiley-Blackwell, Springer, Taylor & Francis, Palgrave Macmillan and Wolters Kluwer are just a few of the prominent publishers who have enrolled their journals as COPE members. Editors and publishers can get guidance from COPE on any topic related to publishing ethics, including how to deal with instances of research and publication misconduct.

Another organization that represents the interests of more than fifty universities and organizations devoted to scientific research is the UK Research Integrity Office. It was founded in 2006 with the following objectives: To provide confidential, unbiased and knowledgeable advice and guidance about the conduct of academic, scientific and medical research; to share best practices for addressing misconduct, poor practice and unethical behaviour; and to promote good governance, management and conduct of research in these areas.

Medical professionals are accountable for protecting the lives and wellbeing of research subjects when conducting scientific research.

One must keep in mind that producing a scientific work necessitates originality as well as openness, honesty, trust and adherence to the ethical guidelines for preparing scientific papers. If the researcher believes that the research could be harmful to the subject if it is continued, then the researcher or the investigating team should stop. When it comes to human study, the interests of society and science should never take precedence over the welfare of the subject. In order to avoid misconducts in research, training researchers in research ethics, what defines research misconduct and the gravity of its consequences is crucial.

Some Important Points

- Research ethics includes guidelines for the conduct and dissemination of research results.
- It includes protection of life, health, dignity, integrity and the right to self-determination of each person/animal.
- It also includes privacy and confidentiality of all personal information of research subjects.
- Biomedical researchers must consider the ethical, legal and regulatory norms and standards for research based on human beings and animals.
- Biomedical researchers must ensure minimum harm to environment.
- Research on human beings and animals must be carried out with full adherence to ethical and scientific norms.
- A physician combining medical research with medical care of any person must ensure that the participation of the patient in no way adversely affects his/her health.
- Research at the cost of risks and burdens of human subjects must be avoided.
- Researchers must constantly assess the risks and costs that human subjects face as a result of their involvement in the study.

- Before the study commences, the research protocol needs to be submitted to the relevant research ethics committee for review, approval and comments.
- Research subjects' privacy and the security of their personal data must be protected at all costs.
- There are ethical duties on publishers, editors, sponsors, writers, and researchers when it comes to publishing and sharing research findings.
- The misconduct in scientific and publishing includes fabricating data and findings; falsifying or altering the results; and plagiarizing, including self-plagiarism, fragmented, repeated and double publishing (duplicate publication), etc.

Suggested Readings

Council for International Organizations of Medical Sciences, *International ethical guidelines for biomedical research involving human subjects*, Geneva, 1993.

Indian Council of Medical Research, *Ethical guidelines for biomedical research on human subjects*, New Delhi, 2006.

Kher, S., 'Informed consent process: protecting subjects rights', in *Basic principles of clinical research and methodology*, 1st ed., edited by S. K. Gupta (ed.), Jaypee Brothers, New Delhi, pp. 93–105, 2007.

Melo-Martin, I., and A. Ho, 'Beyond informed consent: the therapeutic misconception and trust', *Journal of Medical Ethics* 34: 202–205, 2008.

National Commission for the Protection of Human Subjects of Biomedical and Behavioral Research, The Belmont Report, US Government Printing Office, Washington DC, 1979.

Saini, R., and S. Singh, Clinical trials in India: need for bioethics, *The Tribune*, August 9, 2010.

22 Intellectual Property Rights and Research

22.1 Introduction

Intellectual property is a special type of property which is the intangible creation of human intellect. Quite often, research outcome either as a new knowledge or as a solution to a problem leads to creation of intellectual property.

The Liberalization, Privatization and Globalization (LPG) regime emerged in the early nineties after the World Trade Organization (WTO) pushed the importance of intellectual property across the countries of the world.

The research in all branches of learning leads to creation of a number of intellectual properties as continued research paves the way to many technologies, inventions, products, process, etc. Following are the characteristics of intellectual property:

- It is creation of human mind (intellect).
- It is intangible property.
- It has exclusive rights given by statutes.
- It is time-bound and has territorial limitations.

22.2 Intellectual Property Rights

All new outcomes generated through research have scope for patenting under intellectual property rights. The two main categories of intellectual property rights (IPR) are:

- Copyrights and related rights.
- Industrial property rights.

Intellectual property rights protect one's work from being used unfairly by others. All kinds of works including creative works, artistic works like images, symbols, etc., literary works like an article or a book and new inventions like a new drug are protected by intellectual property rights which ensure that the benefits accruing from one's work are protected.

Mission oriented and continued research often paves way for IPR and patenting. Hence a researcher must have basic knowledge of the nature and type of intellectual property and method of patenting.

Copyright: It gives authors of books and articles the ability to protect their works from misuse. Databases, reference works, software for computers, books, technical drawings, figures and images and other items are covered under copyright protection. By giving

DOI: 10.4324/9781003527183-22

copyright protection to one's work, he/she can prevent unauthorized use of the work by other people. Anyone using must seek the required formal permission before using any copyrighted content in their articles or publications. One can use paraphrasing, summarizing, or quoting to ensure that one gives due attribution anytime he/she refers to a book chapter or research paper in order to avoid plagiarism. Remember that plagiarism is a serious crime. The original source must be cited in such published work. A research thesis having original research work of special nature has scope to be published as a book with copy rights.

Industrial Property Rights: The rights related to industrial property which include invention, patents, industrial designs, trademarks, geographical indications, layout designs/topographies integrated circuits, trade secrets, protection of new plant varieties and so on are covered under industrial property rights.

Invention: The invention is a successful technical solution to a technical problem. An invention can be granted a patent if it is new, non-obvious and capable of industrial/commercial applications.

Patents: An innovation, such as a product or a technique that gives a novel approach to a problem or a new technical solution, is given an exclusive right known as a patent. It prevents others to manufacture the patented products without the permission of the original inventor. The period of patents is normally for a span of 20 Years. Continued research paves way for new inventions or process which has scope for commercial applications.

Industrial Design: What distinguishes and enhances a product is its industrial design. These could be 2-D (lines or patterns) or 3-D (the shape or surface of an object) features. One example of industrial design is the shape of your favourite perfume bottle.

Trademark: A trademark is a distinctive symbol used to distinguish a good or service. It could be a single word or a combination of words and numbers. Symbols, 3-D signs, and even drawings can have trademarks. For instance, the trademark Google is well-known. Depending on the amount of protection necessary, the trademark application may be submitted at either the national or regional level.

Geographical Indication: A geographical indication identifies a product as being from a certain area and attributing its quality or reputation to that area. A watch labeled as 'Swiss made' implies that the technical development, assembly, and final inspections, have taken place in Switzerland.

Layout Designs/Topographies of Integrated Circuits: These are three dimensional arrangements of elements forming an integrated circuit intended for manufacturing. These arrangements and ordering of elements follow from electronic function that the integrated circuit is to perform.

Trade Secrets: These are a type of intellectual property which includes formula, practice, process, designs, instruments, patterns, or compilations of information having economic value.

Protection of New Plant Varieties: The protection of plant varieties and Farmers Rights Act in India allows for the creation of an efficient framework for the protection of plant varieties, as well as the freedom for farmers and plant breeders to promote the breeding and cultivation of new plant varieties.

The World Intellectual Property Organization (WIPO) is in charge of overseeing intellectual property rights. The WIPO harmonizes global policy and protects intellectual property worldwide. A researcher frequently looks to previously published work to develop fresh hypotheses or to bolster their conclusions. Therefore, it must be verified that the copyright of the owner or author of the published work (pictures, extracts, figures, data, etc.) is not violated.

Here arises the question as how does one decide whether to publish or patent? The first step is to review the regional IPR laws. IPR regulations differ between nations and regions. For Example, in the US, a concept that has already been published will not be eligible for a patent. Therefore, it is encouraged that researchers apply for a patent before publishing in a publication about their invention. A researcher has one year in the United States, from the time of public disclosure to file a patent. But in Europe, a researcher who has already made their idea public forfeits their opportunity to apply for a patent straight away.

IPR laws can impact international research collaboration. When preparing for international collaboration, researchers should consider the laws of the collaborating country. For instance, before submitting a patent application, researchers from the United States or Japan collaborating with researchers in European Union must consent to limit public disclosure or publication. Publicly aided institutions frequently hold onto their patent ownership in the United States. The best partnership gives everyone the maximum ownership of patent rights possible. Researchers also have the option of engaging with companies that specialize in organizing international research collaborations.

22.3 Conditions for Patentability of Inventions

The basic conditions for patentability of an invention are:

Novelty: A patent can only be issued for an invention if it has never been previously described or revealed anywhere in the public domain. Disclosures about invention made during talks with third parties and on personal websites, scholarly articles, abstracts, and presentations at scientific conferences can be detrimental. Hence the researcher must choose between patenting or publications as both are not possible for any specific invention. In other words, secrecy of the invention is vital for patenting.

Inventive step: Even while an innovation may be new, it may not be thought to have any inventive steps or procedures. An invention should be such that a person working in the same field should not come up with the same invention using common knowledge or by employing information specifically related to that field. It is obvious that this requirement cannot be evaluated as simply as the novelty criteria. In reality, it is a matter of trying to develop arguments that will persuade the patent examiner (who ultimately decides whether a patent is issued or not) of the innovative step. A patent attorney helps in determining if an invention involves an original thought or inventive step.

Industrial applicability: Any invention with practical industrial uses is eligible for patenting. Most of the patent applications pertain to industrial applications.

In addition to the previously specified conditions for patentability, the disclosure of the invention must also meet a number of other formal requirements. The demand for support for the invention is one of these formal conditions. The disclosure must provide sufficient details about the invention so that a person knowledgeable in the relevant field can duplicate it.

22.4 Patentable Inventions

Patentable inventions do not necessarily need to be sophisticated or even intelligent. Throughout their careers, many scientists produce several inventions. To allow the invention to catch the attention of people, researchers must have a particular level of awareness. In general, an innovation can be patented if it is new, not obvious, and has an industrial use. A substance, mixture, formulation, tool, or procedure can all be considered as inventions.

22.5 Grant of Patent

The granting of patent to an inventor by the country-level authorities is done after the strict scrutiny of applications by them. In country like India, objections can be raised by other people during pre-grant and post-grant periods. The patents are applicable within the boundaries of the country granting patent to an invention.

The international organizations dealing with intellectual property include World Intellectual Property Organization and World Trade Organization. The patent office in India is headquartered at Kolkata with branch offices at Delhi, Mumbai and Chennai.

22.6 Licensing of an Intellectual Property

In order to prevent the misuse and also to allow the use of intellectual property of a person by others, there are adequate rules and regulations. Licensing is one such measure through which a person can use the intellectual property of others on mutually agreed terms without compromising on the actual ownership of the IP. It allows the owner of the IP to earn income on one side and on the other side, potential markets are generated for the technologies developed by a particular person for the benefits of the society through the transfer of technologies.

22.7 Licensing Conditions of IPRs

The licensing is permission for commercial use of the invention by others based on the facility available for actual owner of the IP to transfer the technologies through licensing under terms of agreements. It is possible to allow the use of technology partially or fully. The royalty will have to be paid by the user to the owner of IP. The terms may bind the user from transfer to a third party. The royalty payment can be in proportion to the production or sale of products.

22.8 Benefit of Protecting the Intellectual Property

The main incentive for academics to participate in the development of IPR should be the satisfaction that comes from watching a technology that they first invented become a finished good that eventually helps people. In addition, there is the possibility for financial gain, even though there may not be a great likelihood that any individual patent application will result in significant earnings. The number of patents is a matter of pride and recognition for both the organization and the individual researchers.

Academic researchers frequently lack the awareness, business perspective, or in-depth knowledge of IP-related concerns necessary to effectively proceed with patenting their invention and publishing their findings, which is a substantial bottleneck that still exists. The issue here is that a public disclosure of research findings may negate the novelty of the invention, which reduces the likelihood that such invention will ultimately benefit the end-users of their research.

22.9 Research-Based Invention to Patent

The term 'intellectual property rights' is used to describe a wide range of legal rights for different types of intellectual creations. IP rights enable owners to prevent others from commercializing their works. Depending on the nature of IP, rights are either automatically established or require application. A patent application must be submitted in order to protect an 'invention'. A solution to a technical problem can be regarded as an invention. For research organizations, other IPR types like copyright, trademarks, etc. are less important.

The purpose of the patent system is to strike a balance between the needs of society as a whole and the interests of the inventor. While the creator is given 20 years of exclusive use of the invention, the underlying knowledge is made available to the public, allowing others to improve upon the idea and develop new innovations.

22.10 Problems for Patent Protection in Academics

Normally, publishing and sharing research findings to the professional society are a researcher's top priority. This is in contrast to patent protection, which demands utmost uniqueness in an innovation. Scientists must take steps to prevent losing their patentability, by simple practices which would ensure innovation protection while allowing appropriate and timely scientific exchange.

As long as the patent application is filed prior to the publication date, it is feasible to submit abstracts and manuscripts before patenting their content. In fact, during the review process, conference abstracts and manuscripts submitted for publication are typically kept confidential; it is only after their final publication that their content is no longer regarded as innovative. A manuscript that is almost finished is a great foundation for a patent application. A patent attorney can draft and submit a patent application based on a completed manuscript. The time needed for a scientific article to be published is substantially longer than this.

In addition to scientific publications, circumstances that may result in the loss of novelty include discussions of student dissertations, oral presentations, and even scientific conversations with co-workers. When discoveries are revealed at presentations made within the same institution, that is, exclusively to persons who work for the same employer as the inventors, complete secrecy is not always necessary. As long as those in the audience are (made) aware that they should not divulge the substance of the presentation elsewhere, such presentations are not damaging to the uniqueness of an invention.

22.11 Inventorship

The rights of the applicant and inventor are not usually known to scientists. It is crucial to stress that the notion of an 'inventor' may differ from that of a principal investigator. In patent law, a person or group of people who conceptually contributed to the claims detailed in the invention disclosure are considered inventors. Most of the time, one cannot claim to be an innovator just by facilitating the reduction of the invention to practice. For example, a department head who has provided the financing, lab infrastructure, and technicians as well as job and salary to the possible inventor but not contributed intellectually does not qualify for claiming stake in inventorship. This is different from being a co-author of a scientific publication. It is crucial for both political and legal reasons that (possible) inventors thoroughly document all IP-related actions to disprove erroneous claims made by individuals who are not entitled.

However, in most countries, the research institute that employs the inventor is the applicant of patent application by institute job regulations and therefore retains ownership of the patent.

22.11.1 Patent Filing

The amount needed for filing and maintaining patents is typically constrained, particularly within universities. The decision to patent a particular invention is therefore not made lightly and typically entails evaluation of the patentability criteria by a trained patent attorney as well as commercial evaluation of the business case. Some questions that need to be raised before applying for patent include:

- Is the invention actually solving a pertinent problem?
- Is the market for the invention big enough to get a company's attention?
- Is it possible to persuade the company to license the technology in advance and to develop the product later?
- Who are the competitors in the potential market?

Once it is decided to file a patent application, the next step would be to decide on the time to file the application. The likelihood that someone else will come up with the identical innovation and file a patent earlier is reduced by filing as soon as possible. The first person to file a patent on an invention is granted the patent, even if someone else may have come up with the idea first. This is because the patent system is built on the first to file concept. It is better to file for a patent before submitting a journal article or presenting the invention in a conference.

There are a few things that need to be considered before deciding on the time to file a patent application. Firstly, once a patent application has been filed, deadlines that depend on the filing date specially the patent's expiration date need to be fixed. Most development expenditures are incurred at the early stages of a patent lifetime, but majority of the earnings are typically received towards the end. This is a strong justification for submitting the patent as late as possible. Secondly, in case of biomedical patent applications, the scope is limited by the quantity of technical evidence provided. For instance, when a compound is claimed for use in treatment of a new disease, there may be similar compounds which have the same effect. However, in the absence of supporting data, it might be more challenging to acquire a broad protection for an entire class of compounds. Additionally, medical use claim requires in vivo data which demonstrates a compound's effectiveness in a treatment. Thus, the first filing date can be postponed to allow for the collection of such data, giving the institute a better chance that its investment in the patent would result in a profit.

Usually, the patent attorney is in charge of writing the patent application. The writing of the patent application should ideally begin with the formulation of the invention's claims, typically based on the manuscript on the invention which has been prepared to be submitted to a scientific journal. Before filing the patent application, the patent attorney will invite the inventor(s) to review the application a couple of times to respond to queries and offer further input. The entire process can be completed in a few weeks.

22.11.2 Patent Cooperation Treaty Phase

The patent applicants have what is called the 'priority right' to file within one year a subsequent patent application for the same invention. The first filing date is used to assess novelty and innovative step. This right is generally used to file an international patent. An international patent is filed under the Patent Cooperation Treaty (PCT) which is an agreement between more than 160 nations to recognize the priority of each other's first patent application. By filing a PCT application, one can postpone filing separate applications in individual countries for 30 months from the first filing. Due to high cost of PCT application, those applications which do not have sufficient data after the first year are discontinued. Once the application enters PCT phase, additional data providing further support for claims can be added.

22.11.3 National Phase

The applicant must decide which nations/regions the application should be granted in 30 or 31 months (depending on the territory) following the first filing of the patent application or 18 months following the subsequent PCT file. The goal is to get the application approved primarily in those jurisdictions where there is thought to be a sizable market for the invention's product(s) or where the main competitors are situated. This may be highly expensive, depending on the number of jurisdictions elected. The expenses include filing fees, patent attorney fees (a separate intermediate agency is necessary for each international jurisdiction), and translation fees, which are applicable to applications in, for example, Chinese and Japanese.

An examiner from the national patent office further examines the patent application in each jurisdiction after it enters the national phase. It is extremely uncommon for a patent to be issued without any input from a regional examiner. Examiners frequently disagree with the breadth of the claims, which forces the applicant to either develop counter arguments or narrow down the scope of the claims. These exchanges frequently take the form of written correspondence, or 'office actions'. The issuing of the patent is the intended result of these administrative operations or office actions. However, it is possible that all of the claims would be rejected, which would prevent the patent from being granted. Another possibility is that the claims that are ultimately approved are useless from a commercial standpoint, in which case the applicant would have to decide to withdraw the application. A rival or possible patent infringer may also oppose the patent in an effort to have it cancelled, for instance by claiming that the invention should not have been awarded due to a lack of novelty or inventiveness.

The role of the researchers during this phase is limited to assisting the patent attorney in the defence of the claims by responding to any arguments the opponents may have.

22.12 Costs Involved in Filing a Patent

Official filing fees and attorney's fees make up the majority of the costs associated with the first and PCT filings. The costs can vary depending on the application's size (number of pages and illustrations), the nation where it was filed, and the attorney engaged. Translation expenses must be taken into account at the national phase. Using an internal patent attorney can result in much lower costs.

Some Important Points

- The intellectual property types include copyrights and rights related to industrial property which includes patents, industrial designs, trademarks, geographical indications, layout designs/topographies integrated circuits, trade secrets, protection of new plant varieties and so on.
- The research in all branches of learning leads to creation of a number of intellectual properties.
- The intellectual property is creation of human mind (intellect) and intangible.
- IPR is given by statutes, attended with limitations and exceptions, time-bound and territorial and the right is ensured through patenting. A patent is an exclusive right granted for a new invention. An invention can be a substance, a composition a formulation, a device or a method.
- The invention is a successful technical solution to a technical problem. An invention can be granted a patent if it is new (not known to public prior to claim by inventor), non-obvious (invention would not be obvious to a person with ordinary skill of the art) and capable of

industrial/commercial application (invention can be made or used in any useful, practical activity).

- Patents are granted by national patent offices after publication and substantial examination of the applications. In India provisions exist for pre-grant and post-grant opposition by others.
- Licence is a permission granted by an IP owner to another person to use the IP on agreed terms and conditions, while he continues to retain ownership of the IP. Licensing creates an income source for the owner.
- The prime reason that should motivate a researcher to be engaged in the development of IPR is that it is very rewarding to see a technology, originating from a researcher is developed into a final product that ultimately benefits the people.
- A significant bottleneck in patenting is that academic researchers often do not have the awareness, business mindset, or in-depth knowledge of IP-related issues to efficiently proceed with patenting their invention in addition to publishing their data. The problem here is that a public disclosure of research findings may destroy the patentability of any invention arising from data contained in the publication.
- 'Intellectual property rights' is a term used to describe a variety of legal rights for different types of creations of the mind. IP rights provide owners the right to exclude others from commercializing their creations.
- The patent system was put in place to balance the interests of the inventor and society at large, while the inventor is granted 20-year exclusive use of the invention, the underlying information is disclosed to the public.

Suggested Readings

Bindal, S., *Intellectual property law—an introduction*, 2nd ed., EBC Webstore, 2023, https://ebcwebstore.com

Chauhan, A., and K. Singh, 'Intellectual property rights in the digital age: a scopus-based review of research literature', *Journal of Emerging Technologies and Innovative Research (JETIR)* 10(7): 2023.

'Intellectual property rights: an overview and implications', www.ncbi.nim.nih.gov>pmc

'Intellectual property rights: what researchers need to know', https://www.enago.com.academy

Md. Mahfooez Nomani, Z., and F. Rahman, *Intellectual Property Rights (IPRs) and economic development*, New Century Publications, New Delhi, 2018.

Pharmaceutical Research and Manufacturers of America (PhRMA), 'Drug discovery and development: understanding the R & D process', 2007, www.phrma.org/sites/default/files/pdf/rd_brochure_022307.pdf

Rattan, J., *Intellectual property rights*, Volume 1, Bharat Publishers, New Delhi, 2024.

Index

Note: Page numbers in *italics* indicate a figure and page numbers in **bold** indicate a table on the corresponding page.